NO DIG, NO FLY, NO GO

HOW MAPS
RESTRICT
AND CONTROL

Mark
Monmonier

THE UNIVERSITY OF CHICAGO PRESS
Chicago and London

MARK MONMONIER is distinguished professor of geography at Syracuse University's Maxwell School of Citizenship and Public Affairs.

The University of Chicago Press, Chicago 60637
The University of Chicago Press, Ltd., London
© 2010 by Mark Monmonier
All rights reserved. Published 2010
Printed in the United States of America

19 18 17 16 15 14 13 12 11 10   1 2 3 4 5

ISBN-13: 978-0-226-53467-1 (cloth)
ISBN-10: 0-226-53467-7 (cloth)
ISBN-13: 978-0-226-53468-8 (paper)
ISBN-10: 0-226-53468-5 (paper)

Library of Congress Cataloging-in-Publication Data
Monmonier, Mark S.
    No dig, no fly, no go : how maps restrict and control /
Mark Monmonier.
        p.   cm.
    Includes bibliographical references and index.
    ISBN-13: 978-0-226-53467-1 (cloth : alk. paper)
    ISBN-10: 0-226-53467-7 (cloth : alk. paper)
    ISBN-13: 978-0-226-53468-8 (pbk. : alk. paper)
    ISBN-10: 0-226-53468-5 (pbk. : alk. paper)
    1. Cartography—Social aspects. 2. Maps—Social aspects.
3. Cartography—Economic aspects. 4. Maps—Economic
aspects. 5. Cartography—Political aspects. 6. Maps—
Political aspects. I. Title.
    GA108.7.M664 2010
    303.3—dc22

                                                2009033338

The General explained to them clearly and briefly
the object of his visit among them, and, if they were
willing to comply with his demands, the white and
red man could once more be at peace. The country
below Pease creek was shown to them upon the map,
the boundary defined, and if they were disposed to
go there and be at peace, and not cross the line, they
should remain unmolested for the time being.

"Affairs in Florida," *The New-Yorker*, June 22, 1839

A redrawn zoning plan for Troy . . . would allow
residents to keep three dogs or cats, an increase from
the current maximum of two, Davis said. . . . "We felt
that in a lot of cases people might have three indoor
cats. We didn't want to deem them to be criminals."
Zoning violations can cost a homeowner $100 a day
until they are corrected.

Dayton (Ohio) Daily News, September 7, 1999

CONTENTS

Preface and Acknowledgments   xi

1   *Introduction: Boundaries Matter*   1
2   *Keep Off!*   6
3   *Keep Out!*   30
4   *Absentee Landlords*   51
5   *Dividing the Sea*   70
6   *Divide and Govern*   86
7   *Contorted Boundaries, Wasted Votes*   104
8   *Redlining and Greenlining*   117
9   *Growth Management*   130
10   *Vice Squad*   146
11   *No Dig, No Fly, No Go*   160
12   *Electronic Boundaries*   180

Notes   189
Selected Readings for Further Exploration   217
Sources of Illustrations   223
Index   229

PREFACE
AND
ACKNOWLEDGMENTS

Map history can be sliced and diced in diverse ways. To cope with the challenges of language and culture as well as the field's enormous richness, writers typically concentrate on a particular period and country, say, eighteenth-century France. They might also focus on a specific person or problem—Dava Sobel did both in *Longitude,* the saga of John Harrison's heroic effort to perfect a cartographically precise timepiece—or examine a particular map or type of map, for example, the infamous Vinland Map or the maps of exploration and discovery more generally. As an examination of the rise and pervasiveness of what I call prohibitive cartography, *No Dig, No Fly, No Go* reflects the latter approach, but with necessarily varied examples because the story of the restrictive map leads in many directions. I look mostly at twentieth-century North America because that's where much of the development occurred—I say "mostly" and "much" because insightful examples from other regions are unavoidable and roughly parallel developments on other continents are readily apparent. Although a more complete book might justify a decade of additional research, I'm convinced I've captured the essence of prohibitive mapping because cartography worldwide started becoming significantly more homogenized since the middle of the nineteenth century. France and The Netherlands, for instance, are far less cartographically distinct nowadays than several centuries ago.

Insofar as prohibitive cartography works by placing boundary lines on maps, *No Dig, No Fly, No Go* is the latest installment in my series of short integrative cartographic histories that explore the evolution and impact of an important symbol or map feature. The initial volume was *Rhumb Lines and Map Wars: A Social History of the Mercator Projection* (2004), which focuses on the innovative and occasionally controversial grid lines that frame world maps. In a similar vein, *From Squaw Tit to Whorehouse Meadow: How Maps Name, Claim, and Inflame* (2006) delves into the arcane world of standardized place and feature names, while *Coast Lines: How Mapmakers Frame the World and Chart Environmental Change* (2008) is summed up pretty well by its title and subtitle.

My original plan was to write a narrower sequel to *How to Lie with Maps* (1991, 1996), which is a comparatively abstract and timeless exploration of the inevitability and the consequences of map generalization. Although *No Dig, No Fly, No Go* also recognizes the map as a powerful tool of persuasion, the earlier book's heavy reliance on simple hypothetical examples wouldn't work here, largely because deliberate deception or unwitting self-deception is less important to prohibitive cartography than social acquiescence—our unconscious acceptance of cartographic boundaries of all types as natural, beneficial, and worth obeying. To comprehend the rise and broad implications of prohibitive cartography, it's necessary to examine a widely representative range of real maps.

To avoid leading the reader on distracting forays into abstract treatments of power and territory, I focus on showing how prohibitive cartography works as a mapping tool and largely avoid the secondary academic literature. Readers eager for additional information should benefit from the selected readings listed by chapter toward the back of the book.

Because prohibitive cartography is truly pervasive, identifying relevant examples was easier than finding truly insightful exemplars that work well on a small page in black-and-white. I am particularly indebted to Cathy Kittle and Jeremy Hurst at the New York State Department of Environmental Conservation; John Olson and Elizabeth Wallace, respectively the geographic information and earth science librarians at Syracuse University; John Long at the Newberry Library, in Chicago; and Jolai Jenkins, in the mayor's office at Tuskegee, Alabama. Terry Simmons and Karen Culcasi provided useful leads, and George Demko and Lee Schwartz helped nail down the details of travel restrictions on Soviet diplomats and journalists. Joe Stoll, staff cartographer at Syracuse University's department of geography, was a valuable ally in sussing out the complexities of Adobe Photoshop, and the Information and Computing Technology staff at SU's Maxwell School, especially Brian von Knoblauch, Stan Ziemba, Mike Fiorentino, and Ed Godwin, provided rapid response to issues of file storage, network access, and Vista. For broader moral support I am indebted to my colleagues on the History of Cartography Project, particularly Matthew Edney, Mary Pedley, Jude Leimer, and Beth Freundlich. Judy Tyner and Susan Schulten provided helful comments on the penultimate draft. Thank you, too, to Christie Henry, Dmitri Sandbeck, Mark Reschke, Stephanie Hlywak, and Isaac Tobin for the ever reassuring imprint of Chicago style.

CHAPTER ONE

....

INTRODUCTION
BOUNDARIES MATTER

Maps exert power in two ways: by shaping public opinion and by telling us where we can't go and what we can't (or must) do in specific places. This book examines the second type, which I call imperative maps because of their similarity to imperative sentences—the bossy ones that often end in an exclamation point. Whether blatant or subtle, the imperative map is usually intended either to stifle movement or to restrict an activity with a spatial dimension. Examples include aeronautical charts with "no-fly" zones, world political maps, and municipal zoning maps, backed up, respectively, by military aircraft, border guards, and code enforcement officers. The genre also embraces floodplain and fault-zone maps, enforced jointly by environmental agencies and Mother Nature. Whether the penalty for defiance is explicit or implied, an imperative map is a geographic threat that warns of unpleasant consequences. Not surprisingly, most restrictive maps are blatantly prohibitive.

Prohibitive cartography emerged as a distinct dimension of map use sometime after 1900, when restrictive maps increased markedly in variety, pervasiveness, and impact to reflect the growing complexity of cities, governments, and corporations. Although this intensification has roots in Roman property maps, partly intended to thwart trespass, any map with boundary lines delineating a territory as small as a farm or building lot, or as large as a nation-state, is fundamentally a restrictive map. Long-standing public acceptance of property maps and other boundary maps quite likely underlies an expectation by government officials that a wider, more intensive use of prohibitive maps would be understood and accepted. Prohibitive elements are now apparent in most cartographic modes and institutional practices.

Factors underlying this expansion include advances in transportation technology and public administration as well as an increased wariness of urban growth and hazardous environments. While maps portraying historic districts and marine protected areas are necessarily prohibitive, nautical charts and many recreation maps include restrictive elements but largely address other, more important concerns. And even the comparatively nonauthoritarian topographic map shows municipal and state boundaries, which can separate marked differences in tax policy, criminal codes, zoning laws, and environmental regulations.

Perhaps the quintessential prohibitive map is the aeronautical chart, which defines and regulates navigable airspace. Complex and often ephemeral restrictions embedded in contemporary aeronautical charts reflect a historically significant transition from the map as a tool for exploration, discovery, and navigation to the map as a comparatively complex instrument with roles that include public safety, growth management, and environmental protection. Among the diverse roles of prohibitive cartography, the no-fly zone has become a tool of humanitarian intervention, and map-based regulations are indispensable in wildlife conservation

Prohibitive cartography has its own graphic rhetoric. Because efficient enforcement depends on well-defined territorial restrictions, the primary symbol on most prohibitive maps is the boundary line, underscored perhaps by labels and contrasting colors. By convention, small-scale political maps printed in color rely heavily on dissimilar hues—when France is green and Germany is purple, there's less need for the prominent dot-and-dashed "international boundary" symbol common on graytone maps. The mapmaker can emphasize national sovereignty with thick, solid black lines—the most prominent symbol on many State Department maps—or underscore disputed or otherwise tentative boundaries with equally thick dashed lines. By contrast, the thinner, less prominent dot-and-dashed line is a convenient, readily understood code useful with less strident atlas maps, on which political boundaries have a weaker claim to continuity than roads, railways, and rivers. By chance, line symbols with periodic gaps afford a more accurate representation of boundaries like the U.S.-Mexico border, which is far more permeable than an unbroken line might suggest.

Just because a boundary is mapped is no reason to assume it's accepted by the neighbors it separates or by the world at large. Indeed, of all features shown on maps, boundaries are by far the most contentious. While most boundaries signify peaceful sovereignty or undisputed ownership, more than a few assault national pride or personal dignity, often with tragic results. Aggrieved nations declare war, feuding neighbors go to court, and parents upset over reconfigured school attendance zones elect a new school board or pick up and move. Any sudden, unexplained change to a mapped boundary can incite spontaneous resistance, and festering resentment of an old boundary can evoke a forceful reaction.

No less controversial are boundaries tied to zoning ordinances and wetlands regulations. In much the same way that lot lines confirm a property right to sell land and exclude trespassers, boundaries that restrict the owner's use of a parcel transfer some of that property right to the public at large. Imposing these restrictions can be enthusiastically embraced as responsible government or vigorously condemned as an "unconstitutional taking."

Boundary maps have a rich history, with ancient roots and a modern resurgence. Some of the earliest maps no doubt portrayed property boundaries, but the cartographic record here is spotty at best. Because primitive drawings rarely survived, archaeologists can only assume that ancient Egyptians diagrammed the painstaking measurements used to reestablish boundaries after the Nile's annual inundation. Historians trace the modern property map to the Roman Empire, where boundary stones were altars to Terminus, the god of boundaries, and anyone who destroyed or moved one was subject to death or a large fine. Roads, aqueducts, and other engineering works required maps, to be sure, and Roman town plans described the layout of streets and structures. By contrast, national frontiers were comparatively vague and mapped largely by military commanders, whose topographic maps focused on rivers, mountains, settlements, forts, bridges, and roads. National boundaries gained prominence much later, in sixteenth-century atlases, in which recorded borders reinforced notions of national identity, but imperative maps were not a coherent cartographic genre until the twentieth century.

Boundary lore includes fascinating tales of ignorance and greed. It also affords an opportunity to explore popular misconceptions about the accuracy and significance of boundary lines. For example, the perimeter depicted on a property map usually says nothing about the rights of neighbors or utility companies to cross or control part of the parcel. The fundamental document in any real estate transfer is the title, which defines the rights conveyed and provides a legal description of the boundary, typically as a list of monuments, distances, and directions. But nearly as important are easements, rights-of-way, and encroachments, which are poorly understood unless mapped. Boundary law adds its own quirks, most notably on broad floodplains, where a meandering river can slowly cede a landowner's lot to a neighbor.

Far more troubling are international and provincial boundaries drawn up hastily with little concern for inhabitants, resources, and long-term consequences. A classic example is the arrogant partitioning of Africa in the late nineteenth century by European imperialists unaware that putting feuding tribes in the same colony was a recipe for insurrection and genocide. The Nigerian Civil War of the late 1960s and the more recent tragedy in Darfur attest, sadly, to the hidden costs of boundaries designed for the convenience of colonizers and anchored all too easily to a meridian, a parallel, or a river.

Boundary maps have become more sophisticated since satellite positioning made it easy (but not always reliable) to determine whether a car, an airplane, or a known sex-offender was within a polygon defined by a list of latitude-longitude coordinates. At a more local scale, a properly calibrated GPS (global positioning system) receiver can prevent contractors from digging into a gas main or telephone cable—the long, thin buffer zone centered on a pipeline or buried cable is the perhaps the late twentieth century's most noteworthy contribution to electronic boundary technology.

As these examples suggest, this book explores the momentum and impact of prohibitive cartography across a range of scales and phenomena. Chapter 2, which focuses on property boundaries and real estate law, looks at land survey systems and land registration practices, while chapter 3, which deals with national sovereignty, limns the marking and adjustment of international borders, the questionable effectiveness of walls and security fences, and the rhetorical role of boundary maps in asserting spurious claims and fictional sovereignty. Chapter 4 turns to colonial ambitions and geopolitics and appraises the boundaries imposed by imperial powers on Africa, the Balkans, and the Middle East. Chapter 5 looks at offshore and maritime boundaries, while chapter 6 examines the implications in the United States of state, county, municipal, special-district, and tribal boundaries. Law and litigation frame chapters 7 and 8, which treat gerrymandering and redlining, often condemned as subtle (or not so subtle) forms of apartheid. Legal restrictions also figure prominently in chapter 9, which examines zoning and environmental protection, and in chapter 10, which explores the use of maps to regulate behavior deemed offensive or socially harmful. Chapter 11 examines the map's role in protecting air travelers, underground infrastructure, and ethnic minorities like the Kurds in northern Iraq, and chapter 12 looks at looks at satellite tracking, the latest and perhaps most ominous manifestation of prohibitive cartography.

Tailored to lay readers, the book offers valuable insights to anyone who votes, owns a home (or plans to), reads newspapers or their electronic counterparts, peruses atlases, or aspires to be an informed citizen. A second audience consists of geospatial professionals who seek a broader view of their work. And because maps can be enjoyed as well as questioned, the book should appeal to anyone fascinated by maps as legal documents, scientific statements, tools of public administration, or works of art. To appreciate maps, one must understand the symbolic clout of cartographic boundaries.

CHAPTER TWO

....

KEEP OFF!

In the murky world of law school textbooks, land ownership is a property right, which the owner can use or sell and which government can tax. Ownership also implies the right to exclude others from a particular territory, marked perhaps with a barrier or prominent perimeter. Unlike dogs and cats, whose superior sense of smell makes marking territory as easy as, well, you know, humans must rely on walls and chain-link fences, which are more expensive to make but don't need refreshing after a heavy rain. Because walls have aesthetic drawbacks, civil society appreciates boundary surveys, which use words and numbers to describe where a landowner could, if necessary, erect a fence. And because buyers, sellers, and public officials need to visualize the boundary survey, property maps helped create the notion of land ownership and became a distinctive feature of Western cartography.

Although the survey alone might be sufficient to register a title deed at the local courthouse, a property map stamped by a licensed surveyor can reassure a prospective buyer that the house, pool, and other "improvements" lie safely within the parcel's perimeter. Be wary, though, of what the map doesn't show. Because rights-of-way, easements, and "encroachments" from neighboring parcels might hinder use of the property years ahead, it's always wise to ask about these infringements as well as encroachments onto neighboring parcels. Only a fool buys property, especially rural property, without consulting a real estate attorney who knows the area sufficiently well to ask incisive questions.

Purchasing a home in a residential subdivision is comparatively straightforward. The buyer's lawyer explains the sales contract and mortgage agreement after confirming that a title insurance firm has searched for liens and other encumbrances and is willing to guarantee "clear title" to the property. The survey can be relatively straightforward if a subdivision map registered at the local courthouse identifies the parcel with a unique lot number or, in a larger subdivision, by block and lot numbers. It's easy to lay out streets on paper and partition a tract into rectangular lots enclosed by straight-line boundaries with stated lengths. If there are no new improvements, an updated survey and map are comparatively inexpensive, but the surveyor will charge extra for marking the lot's corner points and noting possible encroachments. It's worth a few hundred dollars to verify that a neighbor's fence or deck is not over the line.

This chapter examines the role of maps in land registration and real estate law. It looks first at key differences between the unsystematic surveys used since colonial times in the eastern United States and the more structured, rectangular systems employed farther west. In addition to affecting the work of surveyors past and present, official survey systems leave distinctive imprints on roads, political boundaries, and settlement patterns. The chapter also explores the adverse effects of vague boundary descriptions and lost landmarks; the need for property maps to show rights-of-way, easements, and encroachments; and the technical complexities of riverine boundaries.

Metes and Bounds

Because colonies import their civil codes from the mother country, it's hardly surprising that the eastern states inherited the metes and bounds system of land description from the English, who acquired it from the Babylonians and the Egyptians by way of the Romans. The two-part name indicates a dual approach to boundary description: *metes*, pronounced "meets" and derived from the Latin word for measure, refers to the measured distances and angles used to describe straight-line portions of a property boundary, while *bounds* reflects a preference for physical features like roads, streams, and stone walls. As recorded verbally in the title deed, a property boundary starts at a specific "point of beginning" and runs clockwise or counterclockwise around the perimeter as a series of segments, or "calls," which connect at "corners." The calls may be either metes or bounds, with the adjoining landowner often named for added clarity. Unfortunately for later owners, physical features can deteriorate, migrate down slope, or be relocated, while the land of an "adjoiner" can be sold, subdivided, or willed to an heir with a different surname.

Metes *and* bounds is a misnomer insofar as a boundary description can consist of all metes or all bounds. The latter is potentially troublesome and hardly uncommon insofar as a title deed can be drawn up by a lawyer alone, without a surveyor's services, as Alfred Mulford pointed out in *Boundaries and Landmarks: A Practical Manual,* published in 1912. Mulford showed how to find original boundaries with descriptions like:

Beginning at the Northwest corner of the property to be conveyed where it adjoins the Highway leading from Redfield to Minot's Landing and running thence in an Easterly direction along the land of Clifford Hopson until it comes to the land of David Raup, thence, in a Southerly direction along the land of said David Raup to a large stone at the Southeast corner of the land being conveyed where it joins the land of David Raup and land now or late of Olcott Gates, thence in a Westerly direction along the land now or late of Olcott Gates to the East side of the aforesaid Highway, thence in a Northerly direction along the Easterly side of the aforesaid Highway to the point or place of beginning, Containing within the said bound Thirty-five acres of land be the same more or less.

According to Mulford, even if a surveyor had estimated the acreage with care, a lawyer leery of volunteering too much detail might omit measured distances and directions "for legal reasons."[1] While finding the "large stone" can prove troublesome a century or two later if the deeds of adjoiners Raup and Gates were similarly imprecise, today's surveyor must often contend with a subsequent realignment of the "aforesaid Highway" to promote safety or improve drainage.

Landmarks are the best evidence, observed Mulford, whose experience as a land surveyor on Long Island convinced him that trees and other

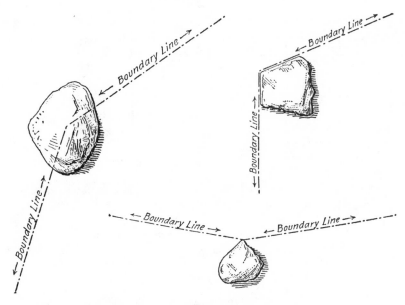

FIGURE 2.1. Examples of the shape and placement of corner stones.

seemingly ephemeral boundary markers seldom disappear completely and that legal descriptions and surveyor's notes, if not followed slavishly, can be useful in finding them. To illustrate what to look for, he included sketches of typical markers and their locations. For example, a boundary stone usually marks the junction of two straight-line calls at a corner point, situated at the center of the stone or indicated more precisely by either a suggestively sharp point or the intersection of two flat faces (fig. 2.1). Rather than anoint all conveniently situated rocks as boundary monuments, the conscientious surveyor looks for stones that "have every appearance of having been placed artificially."[2] No less conspicuous is the tree marked with notches or a stone jammed into its base to indicate a corner (fig. 2.2). Notches on opposite sides several feet off the ground might indicate a "line tree," between corners, but as Mulford warned, "a small boy with a hatchet can mark up more trees in one Saturday afternoon than a dozen surveyors can in a year."[3] Another feature to look for was the "lopped tree," bent or broken as a sapling to align with the boundary. Although he claimed that "such a sapling did not die nor did it ever become erect again,"[4] the nearly severed trunk in Mulford's sketch (fig. 2.3) forecasts death, rot, and ultimate disappearance.

FIGURE 2.2. Notches or a large rock shoved into the roots might indicate a corner tree.

FIGURE 2.3. A lopped tree bent to align with a boundary.

Searching for landmarks can be much easier when the boundary description includes metes as well as bounds, as in this refinement of the all-bounds example:

Beginning at a stake standing at the Northwest corner of the property to be conveyed at a point on the East line of the Highway leading from Redfield to Minot's Landing adjoining land of Clifford Hopson and running thence S 88¾° E, 20.40 chains along land of said Clifford Hopson to land of David Raup as the fence now stands, thence along said Raup's land S 4½° E, 13.04 chains to a large stone at the Southeast corner, thence along land now or late of Olcott Gates S 84° W, 19.11 chains to a large Black Oak tree, thence N 76° W along said lands now or

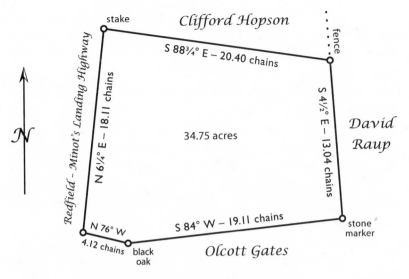

FIGURE 2.4. Graphic description of a metes and bounds survey.

late of Gates, 4.12 chains to the East line of the highway, thence in a Northerly direction along the East side of the said Highway N 6¼° E, 18.11 chains to the point or place of beginning, Containing within the said bounds Thirty-four acres and 120 Square Rods of land, be the same more or less, according to a survey made by Gordon Wolman, County Surveyor, February 2nd, 1817.

By convention, directions like the initial bearing S 88¾° E are referenced first to a line running north (N) or south (S) followed by the angular departure to the east (E) or west (W) of a straight line running to the next corner. As late as the 1970s, surveyors measured distances incrementally using a chain with a hundred identical links and an overall length of 66 feet. This standard length is closely tied to traditional English units for length and area: 80 chains equals a statute mile, 25 links (one-quarter chain) represents a rod, and a rectangle one chain wide and 10 chains long contains an acre, or 160 square rods. In my example, the initial run of 20.40 chains would have been estimated by stretching out the chain 21 times, and counting off only 40 links before the twenty-first increment ended at the fence. I had no difficulty converting this verbal description to its graphic equivalent in figure 2.4. The county surveyor might have used a similar diagram to estimate the parcel's area, and his present-day counterpart would find one handy in searching for surviving landmarks.

A surveyor looking for ancient landmarks won't take an eighteenth-century metes-and-bounds description too literally. Metal expands when heated, making a 66-foot chain a bit longer in August than in January—long enough to account for a discrepancy of a link or two along a half-mile boundary. More troublesome are property lines that run uphill. Some surveyors tried to hold the chain taut and more or less level; others draped it along the ground. According to Mulford, a compromise of sort was common, but "if the country were rough, we are practically sure that the chain was considerably off the horizontal a large part of the time."[5] What's more, a boundary along a stone wall or overgrown fence would have been measured along a roughly parallel line offset a convenient (but perhaps not rigorously consistent) distance to one side.

Questionable distances are not the only problems. The map's north arrow as well as bearings like S 88¾° E, referenced to the cardinal directions, are troublesome because magnetic north (aligned with an unbiased compass needle) seldom coincides exactly with true north (along a meridian). This discrepancy, called "magnetic declination," occurs because Earth's magnetic field is not aligned with its axis of rotation, which explains oddities like the street grid in older parts of Baltimore, Maryland, laid out by compass in the late eighteenth century and about three degrees out of alignment with the meridians and parallels. In addition, magnetic declination varies over time because the north magnetic pole (now residing in northern Canada) is migrating slowly westward, toward Siberia. Although tables and maps describing the planet's shifting magnetic field can help surveyors amend ancient bearings, geophysical data cannot correct for other flaws in angular measurements. Not only was the antique surveyor's compass a crude instrument—its angles could be read only to the nearest quarter degree—but the effects of local attractions such as a nearby telegraph line or rock formation rich in iron were often unknown or ignored. And the surveyor might have read the compass up close, without removing his steel-framed spectacles.

Metes and bounds was the dominant land survey system in the eastern United States, but its impact varies because of regional differences (fig. 2.5) during colonial times in the granting and selling of land.[6] In the six New England states land was typically granted to communities that became the region's present-day towns. Before subdividing their land, they established clear boundaries with neighboring towns, thereby avoiding the overlapping claims and awkward strips of "no man's land" typical of

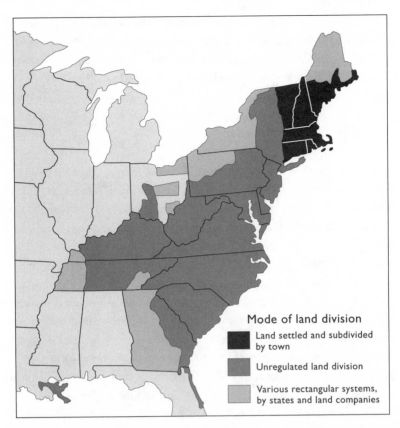

Mode of land division

■ Land settled and subdivided by town

■ Unregulated land division

□ Various rectangular systems, by states and land companies

FIGURE 2.5. The metes and bounds region of the East reflects three distinct types of land division, which Department of Agriculture land-use guru Francis Marschner attributed to administrative practices during the time of principal settlement. The unregulated area, which covers much of the Southeast, also includes Texas, admitted to the union in 1845, after nine years as an independent republic.

the indiscriminate settlement common farther south, where colonial land division was largely unregulated. In Virginia, for instance, a settler might be granted 400 acres without first having his tract surveyed. In areas immediately west, which were settled later, in the immediate postcolonial decades, land companies or state governments set up rectangular grids of boundary lines usually running north–south or east–west. Typically a land company that purchased a tract encompassing hundreds of square miles would hire a surveyor to lay out rectangular lots, perhaps a quarter mile on a side. Arranged in a checkerboard pattern, these lots were numbered for ready identification on a map known as a plat. Straight traverses were comparatively easy to follow across flat or gently sloping terrain, and the predetermined separation of lot lines made it easy to calculate lot size and sell land as a commodity to settlers and speculators. Straight lines were also used to subdivide large parcels into small farms or village lots.

Township, Range, and Section

Still farther west and covering most of the country, a survey system attributed to Thomas Jefferson, himself a land surveyor, was formalized by the Continental Congress in the Land Ordinance of 1785.[7] Eager to raise revenue for the federal government by selling public lands in the new nation's "Western Territory," lawmakers called for "qualified" surveyors to lay out grids of square "townships," six miles on a side and subdivided into 36 square lots, later called "sections," each a mile on a side and containing 640 acres. In addition to running north-south lines a mile apart and marking section corners, the surveyor was to describe each township with a plat showing lot numbers and noting "at their proper distances, all mines, salt springs, salt licks and mill seats, that shall come to his knowledge, and all water courses, mountains and other remarkable and permanent things, over and near which such lines shall pass, and also the quality of the lands." Lot 16 was reserved "for the maintenance of public schools," and lots 8, 11, 26, and 29 were set aside for the federal government. Section numbers start in the northeast corner of the township, increase westward along the top row, drop down to the next row, and snake back and forth in the manner of a farmer plowing a field (fig. 2.6, *upper left*). As an added convenience, straight-line boundaries in a cardinal direction can subdivide a section into halves and quarters—a quarter section was the basic grant under the Homestead Act of 1862—and even into eighths and sixteenths (fig. 2.6, *upper right*). For example, SE ¼ NW ¼ Sec. 14 identifies a square 40-acre parcel, also called a quarter-quarter section, carved out of the 160-acre quarter section identified as NW ¼ Sec. 14.

The rectangular Public Land Survey grid is convenient for identifying townships. North-south boundaries called range lines are aligned with a principal meridian, while east-west boundaries called township lines are laid out parallel to a base line, perpendicular to the principal meridian at the grid's "initial point" (fig. 2.6, *bottom*). Rows of townships running east–west are numbered according to their position north or south of the

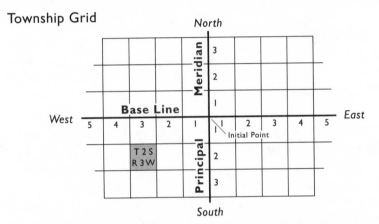

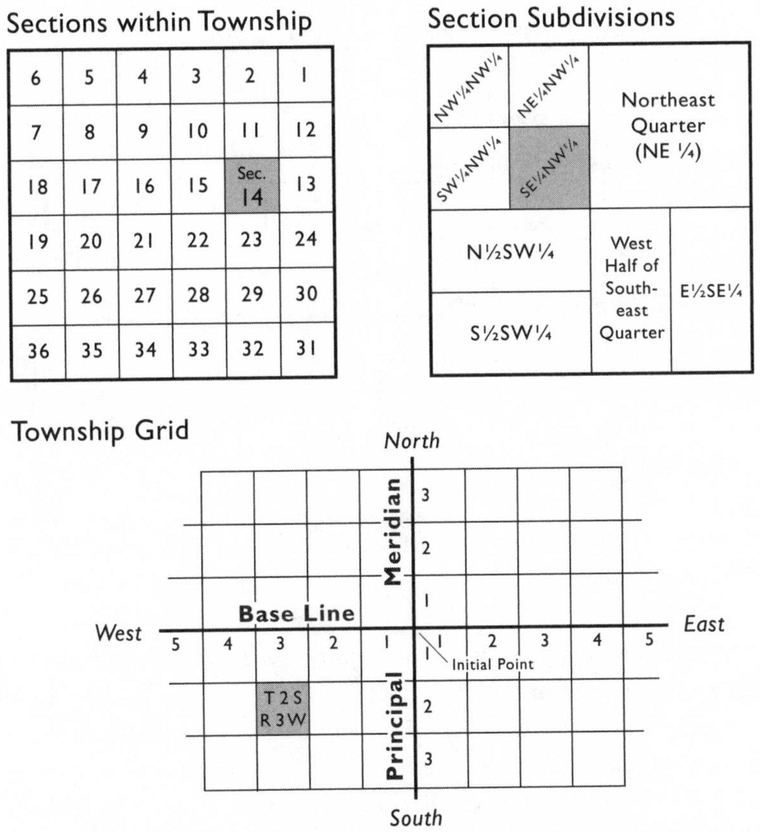

Sections within Township

6	5	4	3	2	1
7	8	9	10	11	12
18	17	16	15	Sec. 14	13
19	20	21	22	23	24
25	26	27	28	29	30
36	35	34	33	32	31

Section Subdivisions

NW¼NW¼ | NE¼NW¼ | Northeast Quarter (NE ¼)
SW¼NW¼ | SE¼NW¼ |
N½SW¼ | West Half of South-east Quarter | E½SE¼
S½SW¼ | Quarter |

Township Grid

North

West — Base Line — East

Meridian — 3, 2, 1

5 4 3 2 1 | Initial Point | 1 2 3 4 5

T 2 S R 3 W

Principal — 2, 3

South

FIGURE 2.6. Key elements of the Public Land Survey grid. Shading highlights the quarter-quarter lot SE ¼ NW ¼, Section 14, Township 3 South, Range 3 West.

base line, while rows running north–south are referenced by their position east or west of the principal meridian. On a map the horizontal rows are called townships, the vertical rows are dubbed ranges, and individual townships are identified by a pair of numbers, for example, T 2 S, R 3 W. It's a straightforward system once you accept the ambiguity of "township" as both a unit of land and a grid coordinate.

The system is a bit more complicated because meridians that are widely spaced along the equator converge, slowly but steadily, to a point at the North Pole. This convergence requires periodic adjustments to keep quarter sections from growing progressively larger than 160 acres with increasing distance south of the base line and ever smaller with increasing distance to the north. To minimize inequality in size, range lines are repositioned

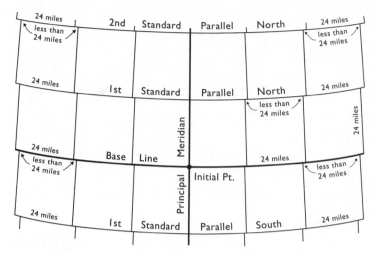

FIGURE 2.7. Offsets along standard parallels correct for converging meridians, which otherwise would produce noticeable variation in the size of section and quarter-section lots.

every 24 miles along east-west correction lines called standard parallels (fig. 2.7). Along the first standard parallel north, situated 24 miles north of the base line, range lines are offset outward from the principal meridian to maintain an exact six-mile separation. A similar adjustment occurs along the second standard parallel north, and the offsets repeat every 24 miles. An exact six-mile separation along the first standard parallel south yields similar offsets along the base line itself. Because these adjustments are cumulative, the offsets along a correction line are more prominent with increasing distance from the principal meridian.

The Public Land Survey had a pronounced effect on the landscape of the central and western states, where neighboring farmsteads were roughly a half-mile apart and roads typically followed section lines. As a consequence of corrections for converging meridians, drivers headed due north or south along a monotonously straight road periodically encounter a sharp, right-angle jog—unless a safety-minded highway department has reduced the danger by blending the two portions with a smooth S-shaped curve (fig. 2.8). The rectangular survey also affected the layout of fields. On clear days air travelers can see a massive checkerboard reflecting differences in cropland and pasture, although in semiarid areas the rectangular field boundaries are often less apparent than an alignment of comparatively lush

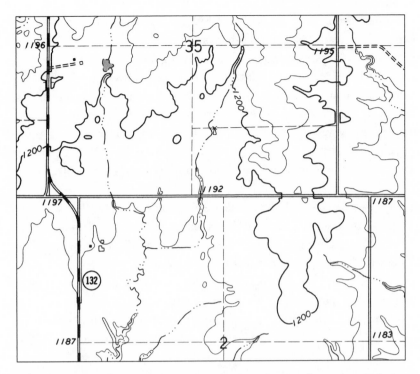

FIGURE 2.8. Offset of section-line roads along a standard parallel. The north–south roads are one mile apart, dashed lines are quarter-section boundaries, and 35 and 2 are section numbers.

circles a half-mile in diameter, formed by central-pivot irrigation systems. Rectangular survey lines are similarly embedded in political maps, on which straight-line town and county boundaries follow cardinal directions.

The General Land Office (GLO), a Treasury Department bureau established in 1812 to oversee development of the public lands, minimized offsets and related distortions by dividing the western two-thirds of the country into 31 zones, each with its own base line and principal meridian.[8] A few zones are small, but most encompass vast areas; for example, a single base line and principal meridian serves all of Nebraska and Kansas as well as most of Wyoming and Colorado, and another zone covers all of Nevada and more than two-thirds of California. In addition, Alaska has five zones, and Ohio has eight more, set up without an official initial point between 1785 and 1805, when federal officials were still figuring how to number townships and cope with converging meridians.[9] To promote standardization and accuracy, the GLO developed a *Manual of Surveying Instructions*, with explicit guidance on running survey lines, keeping detailed field notes, establishing corners, placing monuments, and preparing plats. In addition to establishing section corners, surveyors were required to establish and mark "quarter corners" midway between section corners, to facilitate the subdivision of sections into 160-acre parcels.

Accuracy was a challenge, especially in the early years when the government relied largely upon contract surveyors paid by the mile. According to a 1920 GLO report, shoddy surveying was only part of the problem:

Imperfect methods and instruments were used in earlier surveys. Errors were occasionally made in both the field work and in the office record; corner monuments decayed, and in some cases were willfully removed, but, more regrettable than any of these, is the fact that in some instances frauds were perpetrated upon the government by some of the earlier contracting deputies whose work did not extend beyond the making of a paper record upon which their accounts were paid. Then, too often, the earlier settlers disregarded the official surveys and erected their fences and made other improvements in a careless manner.[10]

Resurveys were carried out as early as 1842, but Congress never provided the money for proper maintenance. And once a survey was officially approved, or "patented," its lines and corners became legal boundaries that created the parcels they enclosed. With flawed lot lines thus frozen in place, reliable plats describing the location of legally established corners became particularly important.

Although an approved survey is presumed accurate, monuments marking a parcel's corners need verification. The landowner eager to subdivide a parcel typically hires a surveyor to examine field notes from the original survey, determine whether monuments that purport to mark the legal corners are consistent with the field notes, and find the correct location of any corner with its monument missing or seemingly disturbed. A retracement survey, as it's called, requires the surveyor to play detective, seek out all relevant evidence, and not be surprised to find that the sides of sections are not 80 chains long, quarter corners are not midway along a straight line between section corners, and quarter sections do not contain exactly 160 acres. An ancient plat with straight lines and right-angle intersections might look authentic, but it's no assurance that the original measurements were accurate or the monuments correctly positioned.

Residential Lots

Maps are essential when real estate entrepreneurs carve farmland into residential lots. The typical developer hires a landscape architect to plan a visually attractive and profitable layout of roads and lots that satisfies state environmental regulations and municipal zoning laws. Mapping is important in experimenting with lot configurations as well as in selling the concept to the local planning board and individual lots to builders or future residents. Because the landscape architect must work within a tract's established boundaries, the savvy developer starts by hiring a surveyor to retrace and map the lines and corners of the quarter section or metes-and-bounds property to be subdivided.

To lessen the likelihood of costly litigation, the developer also hires a lawyer well versed in the principles of "simultaneously created boundaries."[11] Because the subdivision is described by a single plat showing all lot lines at once, the resulting land parcels have equal legal standing. The plat not only represents the seller's intent but can also be used to recover measurements not otherwise recorded.[12] For example, when neither the plat creating a subdivision nor any associated document specifies a street width and the stated scale of the plat is "1 inch = 200 feet," parallel lines one-tenth of an inch apart will establish the intended width of the right-of-way as 20 feet.[13] Moreover, when there are no relevant measurements on the drawing or in the original surveyor's notes, and no relevant physical features to be measured in the field, the plat becomes the record of last resort.[14] Leery of someone measuring the lengths of cartographic boundaries and their angles of intersection a hundred years hence, the cautious surveyor anticipates imprecise drafting and paper shrinkage by setting out durable monuments and writing key measurements directly on the plat.

Larger subdivisions, especially in neighborhoods with parallel streets and right-angle intersections, are usually based on the "lot and block" system, whereby rectangular street lines partition the tract into blocks, subdivided in turn into parcels numbered consecutively within each block. As shown in figure 2.9, unique combinations of block and lot numbers identify individual parcels. Numbered lots can vary in size and shape. While in older subdivisions the lots are usually rectangular with equal dimensions, in subdivisions with winding roads and culs-de-sac it's not unusual for lots to have more than four sides—some straight, others curved—with their

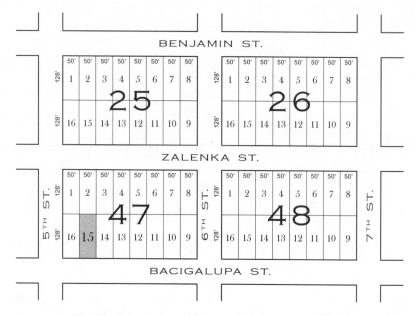

FIGURE 2.9. Hypothetical plat showing key elements of lot and block subdivision. Shaded area is Block 47, Lot 15.

intended lengths noted on the plat. For a lot bounded by a circular arc, the plat typically indicates not only distance along the arc but also either the degree of curvature or the center and radius of the circle. Because of a greater potential for error, a lot with curved boundaries or a highly irregular shape is more expensive to survey and more likely to invite litigation than a lot with simple rectangular borders.

Property owners are occasionally surprised when precise measurement reveals that a block is larger or smaller than the sum of its constituent lots as described by the plat. For example, in laying out the subdivision in figure 2.9, the original surveyor might have established the south side of block 47 by driving iron pipes or pins into the ground only at the block's corners, that is, at the southeast corner of lot 8 and the southwest corner of lot 16. Because the intervening eight lots are all 50 feet wide, it's efficient to place monuments only at the block's corners, assumed to be 400 feet apart, and to apportion the length equally after the lots are sold. This practice is called "subdivision by protraction." If a later, more accurate survey discovers that the block's original corners are really 401 feet apart, the excess is distributed proportionally among the eight lots, each now an inch and a half wider than originally thought. Similarly, if precise measurement reveals an actual distance of 399 feet between the monuments, the deficiency is absorbed equally by the eight lots. Because the original survey establishes each block as a separate unit, a surplus in one block cannot help correct a deficiency in an adjoining block.

Sometimes a tract is subdivided by laying out block boundaries along street centerlines and establishing block corners in the middle of future intersections. A quick and inexpensive way to create a subdivision, it's potentially troublesome decades later when a landowner wants to erect a fence or the city needs to upgrade its storm sewers. To avoid uncertainty over the location of monuments, the city engineer's office might place "offset monuments" beyond the curb line as a substitute for original monuments that are buried under the roadway and easily disturbed by road paving or repairs to gas, water, or sewer lines. A network of these monuments is useful to surveyors who want to locate the corners of residential lots as well as to engineers concerned with maintaining roads and underground utilities. Although offset monuments do not carry the clout of original corners, they are presumed correct if accepted without dispute for a long period.

Durability and protection are essential. In Syracuse, New York, where I live, the typical municipal monument is a granite block four feet long and six inches square. Placed vertically in the ground with its upper end about 10 inches below grade, it's resistant to frost heave and careless contractors. A quarter-inch hole drilled in the center of the top end marks the survey point, protected by a cylindrical metal "crock" covered at grade level by a removable metal plate labeled "Mon." Maps and original field notes archived at the city engineer's office relate each monument's location to earlier surveys, prominent landmarks, and other monuments. The monument network was started around 1904, extended in response to several annexations, and resurveyed in the 1980s using modern electronic equipment. Vandals have removed a few of the crock plates, but mischievous kids are less troublesome than the idiot with a backhoe.

Another municipal department concerned with survey lines and property boundaries is the assessor's office, sometimes called the Real Property Bureau. The name varies from county to county and town to town, but the goal is similar: identify all parcels and their owners, estimate the value of land and structures, and generate tax bills. Although local officials responsible for taxing real property have always relied on land records of some sort, in the final decades of the twentieth century, they began to use computers and aerial photos to compile maps and databases describing all land parcels within their jurisdiction. A systematic inventory of real property seemed an efficient way of making certain that acreage measurements were reliable and that no privately held land fell through the cracks—any tract with no apparent owner might be claimed by the state, sold, and put back on the tax rolls. By researching title deeds and plats, tax officials were able to plot lot boundaries on large-scale, geometrically correct photomaps and identify swimming pools, outbuildings, and other "improvements" that might have escaped the assessor's occasional door-to-door canvass or "windshield survey."

Modern tax mapping is potentially more equitable than older, file-card or book-based systems with maps that merely record owners' names. Municipalities with a "transfer tax" based on sale price can implement a "full-value assessment" program based on what recent buyers paid for similar nearby parcels. Although a buyer who overpaid might aggravate neighbors who suddenly find their assessments increased, full-value assessment is eminently fairer than the "soak the newcomer" approach common where assessments are not updated regularly. As public documents, tax maps empower homeowners to challenge their assessments on "grievance day" by pointing out clearly more valuable nearby properties with lower tax bills. And because assessments and sale prices are readily available—many jurisdictions put their tax maps online—the local real estate market benefits from more knowledgeable buyers and sellers. While tax maps lack the legal standing of title deeds and surveyor's reports, they put property boundary data in the hands of other interested users, including urban planners, marketing analysts, and environmental scientists.

Canada, Australia, and Western Europe, which are well ahead of the United States in showing accurately surveyed boundaries on large-scale "cadastral maps," are moving toward electronic land-registration systems that represent land parcels as polygons and corners as pairs of (X, Y) plane coordinates. When these systems eventually replace deeds and monuments, buying land will be no more complicated than buying a used car. Moreover, the homeowner with a highly precise handheld GPS (global positioning system) unit can quickly determine whether a fencepost or rosebush is on her own property or a neighbor's. Converting existing legal descriptions into a massive digital map will not be easy or cheap, but the benefits of simpler transactions, fewer lawsuits, more accurate land valuation, and more effective environmental management can justify the investment.

Easements and Encroachments

An easement is a landowner's legal right to use all or part of another's land for a specific purpose. Legal lingo distinguishes between a *gross easement* like the right-of-way for a telephone cable or municipal sewer and an *appurtenant easement* that enhances a specific property. In the latter case, the property receiving the benefit is the *dominant tenement* and the one encumbered is the *servient tenement*. An easement may be either positive, as when the owner of a "landlocked" lot enjoys "ingress and egress" across a neighboring parcel, or negative, as when a landowner grants a "scenic easement" by agreeing not to block a neighbor's view with trees or structures. Maps are useful because an easement typically affects only a specific part of the servient parcel, whose owner might be sued if the rights invested in the dominant parcel are infringed by a fence or structure—imagine receiving a bill from a utility company for having to blast through concrete when it repaired the gas line running beneath your swimming pool, now no longer useable. For this reason, the surveyor should look for and note all physical evidence of easements, such as driveways, drains, and utility lines. Similarly, the landowner enjoying an easement needs to know its bounds, especially if the easement is geometrically complex. Cartographic description is particularly important for a formal easement that is transferred automatically when either parcel is sold. Laws governing easements vary significantly from state to state, and buyers should also be wary of informal easements that are extinguished through nonuse or cancelled when the servient tenement is sold.

By contrast, an encroachment is a deliberate or inadvertent intrusion onto a neighboring property. It can be as innocuous as a telephone wire strung directly from a utility pole to a house on a neighboring lot, as mundane as a fence a few inches out of bounds, or as troublesome as a six-story office building partly over the property line because of a contractor's mistake. Although the surveyor should look for and plot all evidence of encroachments, including overhanging roofs, fire escapes, and windows that open outward, he should not only reject a request to show "all" potential easements or encroachments but also refuse to assess the legality of intrusions he observes—that's your attorney's job. Encroachment is a legal concept, not a surveyor's term.[15]

An apparent encroachment is worrisome because it could become a *prescriptive easement* if, for instance, a judge finds that your neighbor has a permanent right to leave his hedge or carport where it's been for the past 20 years. (Length of time and other conditions for prescriptive easements vary from state to state.) Worse yet, the judge might view an encroachment as *adverse possession* and award the neighbor part of your land if its use is found to have been exclusive, continuous, "open and notorious," and "hostile"—language that suggests a need to consult a lawyer familiar with real estate law and court rulings in your state. She might advise you to sell the contested strip, negotiate an agreed boundary with the neighbor and file it at the courthouse, or officially grant permission for the encroachment, perhaps with the neighbor paying a small fee and most certainly with a signed agreement or a postal certificate confirming delivery of your letter of permission. Formal permission can keep the neighbor content while fending off a later claim to a prescriptive easement or adverse possession.

When attempts at settlement fail, the ensuing trial often challenges the court to resolve an intriguing mix of conflicting claims and disputed facts. My favorite, because it involves a map and a David and Goliath cast, is *Enos v. Casey Mountain,* decided by Florida's Fifth District Court of Appeal in 1988.[16] The unanimous opinion of the three-judge panel included a map (fig. 2.10) showing the parcel of land (*lower left*) near Orlando bought by Thomas Enos and his wife in 1966 but separated from State Road 535 by the Grand Cypress resort built by Casey Mountain, Inc., a decade and a half later. When the Enoses purchased the property, they assumed guaranteed access to the state highway along the 15-foot wide "Munger Right of Way," shown as a road on a subdivision plat recorded in 1912 and represented on the court's map by a double-dashed line. The subdivision was never surveyed and its roads were never built, but a dirt road along the platted right-of-way served the Enoses well until Casey built its 750-room hotel and nearby conference center. The firm closed the Munger road but let the Enoses use its paved access road marked by arrows around the south and eastern periphery of the complex.

Magnanimous perhaps, but not what the courts consider free and unimpeded access. Irked by having to pass through the Grand Cypress security gate and submit to alleged harassment from Casey's security guards, the Enoses sued to regain their implied easement, no longer linked to the subdivision plat because new evidence revealed their dirt road was situated slightly east of the Munger Right of Way. Casey claimed the Enoses' easement was either extinguished by nonuse (they obviously didn't use

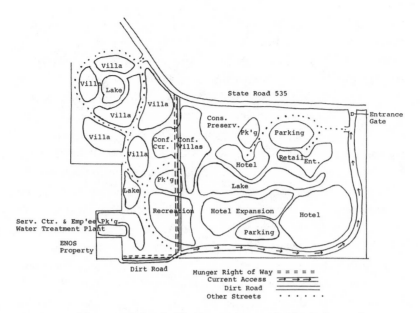

FIGURE 2.10. Map appended to the Florida appeals court decision in *Enos v. Casey Mountain, Inc.*

the dirt road during construction) or by adverse possession (its conference center now blocked the right-of-way). Although a lower court had concluded that the Enoses were adequately served by a mere "license" to use Casey's access road, the appeals court ruled that the couple was entitled to a "substituted or redelineated implied easement" along the route shown on the map. What's more, "no gates or security guards shall limit this easement (which shall 'run with the land')"—in other words, a new owner will acquire the same right.[17] In a decision Solomon would have admired, the Enoses won permanent and unfettered access to the public highway, while Casey kept its conference center but had to pay court costs and the couple's legal fees.

Water Boundaries

Property lines along a riverbank or seacoast are particularly troublesome because natural processes make water boundaries "ambulatory" and state statutes must rely on formal definitions and rules to balance public and private rights. Some laws deal with erosion, deposition, and rising or falling water levels; some address passage rights in navigable waters; and some guarantee access to the public beach, divided by the tides into a "wet beach" between the "low-water" and "high-water" lines and a "dry beach" above the high-water line.[18] In New York, Louisiana, Washington, and Hawaii, for instance, the public not only owns the wet beach but enjoys the right to use the dry beach, reaching inland as far as the "vegetation line." By contrast, Massachusetts is one of five states in which private land extends seaward to the low-water line, with limited public access to tidal lands for fishing, hunting, and navigation. In addition, state statutes draw important distinctions between tidal and nontidal riverbanks, and between navigable and nonnavigable inland waters. Along navigable rivers and lakes, which must be shared with the public, landowners in most states see their rights stop at the "ordinary high-water mark," whereas the beds of nonnavigable inland waters are commonly divvied up among adjoining lots. Generalizations are dicey, though, because the states differ markedly on *riparian rights,* which involve uses as diverse as fishing, building docks, and withdrawing water for irrigation.[19]

Riparian law distinguishes normal erosion and deposition, which are assumed to progress slowly, grain by grain, from *avulsion,* which occurs suddenly and catastrophically, as when a river in flood cuts off a meander loop, the ends of which fill with sand and silt to form an "oxbow lake." According to the principle of avulsion, boundaries stay where they were— state boundaries as well as property boundaries. By contrast, erosion slowly moves the riverbank toward a property's interior and thus diminishes its size, whereas *accretion,* the legal term for slow deposition of sediment along the bank, pushes the boundary outward, increasing the acreage of riverfront lots. Unless an interested party makes a convincing case for avulsion, some states presume any change in water boundaries is the result of erosion and accretion.

A state's legislature and courts can make a big difference in how land formed by accretion is allocated to adjoining lots. Figure 2.11 describes two common methods for apportioning accreted land. Some states extend property boundaries outward from the original riverbank along lines perpendicular to centerline of the river; others apportion the new riverfront in accord with the ratio of frontage along the original shoreline. While most states and the federal government prefer proportionate allocation,

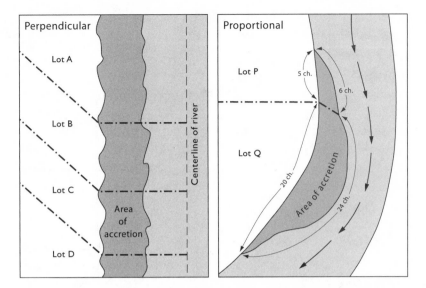

FIGURE 2.11. Perpendicular (*left*) and proportional (*right*) methods for dividing riparian accretions among adjoining lots.

the perpendicular method is useful when the streambed belongs to riparian owners.

When a river has occupied various positions after the original survey, maps showing the channel's position at specific dates can be decisive in working out who owns what—maps like figure 2.12, which the Supreme Court of Kansas included with its decision in a lawsuit filed by Charles Peuker, who owned the shaded lot. The map describes three positions of the northeast bank of the Missouri River: its "present" location in 1900, when the case was argued; its position in 1855, when the GLO surveyed the area; and its position around 1870, when the river had drifted farther to the northeast and eroded part of Peuker's quarter-quarter lot. Over the next 30 years the river shifted well to the southwest, creating new alluvial lands along its northeast bank. The defendants, William and Ella Canter, who owned lots to the south and west of Peuker, were reluctant to concede any of the accreted land despite Peuker's argument that the river's position around 1870 conferred a riparian right to share in any subsequent accretion. The court agreed, and awarded Peuker a proportionate share of the accreted land—more than enough to compensate for past erosion within his lot's original square boundary, as surveyed in 1855.[20]

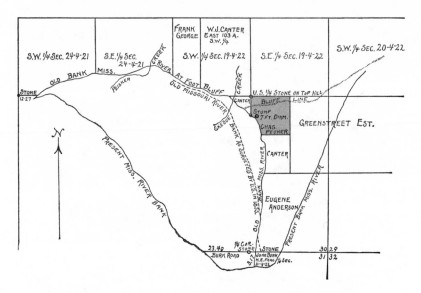

FIGURE 2.12. Map presented as evidence by Charles Peuker and included with the Kansas Supreme Court's decision in *Peuker v. Canter*. Shading added to highlight Peuker's lot.

The *Peuker* decision confirmed that erosion can create a new riparian right for a previously "remote" lot. More ominous is the possibility that a channel moving in one direction might extinguish a landowner's property rights entirely, including riparian rights, but later move in the opposite direction across the original survey lines to accrete land for a neighbor. Some states allow this, others don't.[21]

Because ambulatory boundaries are difficult to survey—measuring an irregular curve is seldom straightforward, and the surveyor could slip and drown—riverfront and shoreline boundaries are usually approximated by a series of straight courses known as a *meander line*. Thought convenient for estimating acreage, the meander line is seldom considered the true boundary.[22] Although the "upland owner" is generally assumed to own the land down to the water's edge, a prospective buyer needs to be certain that the meander line is not the official property boundary; otherwise the lot is considered remote, rather than riparian, and its owner can't build a dock or walk to the riverbank without permission. Because a riparian lot's previous owner might have been neighborly, a prospective purchaser must be wary of easements attached to remote lots. Access easements, common in coastal areas with an attractive public beach, can diminish privacy and land value as well as interfere with intended improvements.

....

As the examples in this chapter demonstrate, maps and surveys share diverse roles in the buying, owning, selling, and subdividing of land, and in thwarting trespass. While the survey that created a parcel might enjoy more clout in court, maps have told surveyors where to run their lines, helped their successors find forgotten monuments and fix lost corners, and left a lasting impact on road networks and settlement patterns. Moreover, by making land ownership legible, maps have promoted the orderly settlement of appropriated territory, given local government a reliable tax base, and formalized the easements and rights-of-way essential for commerce and neighborly cooperation. And when owners of lots or easements need to assert their rights in court, maps provide an efficient framework for organizing evidence and settling disputes.

CHAPTER THREE

....

KEEP OUT!

Maps of national territory have much in common with real-property maps but at a broader scale. As a tool for asserting territorial rights, maps can promote peace between neighboring countries as well as civility between adjoining landowners. And because boundaries are barriers to movement, prohibitive cartography becomes a rhetorical tool for restricting immigration, interdicting contraband, regulating trade, and sheltering money in much the same way that property maps ward off trespassers and mapped easements sanction underground gas lines. Even so, international boundaries are markedly different from lot lines, which lack the instant recognition of tiny maps on postage stamps and the careful scrutiny of military planners and the World Court.[1]

Cartographic Insults

That small, fleeting boundary maps can trigger massive reactions surprises many Americans. Why, they ask, would the Indian government ban the Windows 95 operating system because of a world map displayed briefly during installation? Although the map merely asked users to indicate their time zone, its international borders were deemed much too generous to China and Pakistan, which controlled parts of Kashmir claimed by India. At issue were eight of the map's 800,000 pixels—a tiny fraction of the cartographic stage but a huge expense for Microsoft, which recalled 200,000 copies of the offending software.[2] That anyone noticed an eight-pixel discrepancy is hardly surprising: India is the world's second most populous country, and its deeply-felt claim to Kashmir had prompted previous embargos of politically unacceptable maps and atlases.

Make no mistake: a simple map of national boundaries can be a powerful symbol, imprinted in the minds of citizens and public officials, who react quickly to aberrant representations, however minor or unintentional. Microsoft thought it was acting properly in basing its borders on a UN map but lacked the space for a caveat advising users of the territorial dispute. The firm suffered a similar reaction from Peru over its portrayal of that country's border with Ecuador.[3] Because no single world map with international boundaries can please everyone, Windows now merely asks users to choose a time zone from a list of major cities. The Date and Time Properties menu in the control panel on my laptop displays a small, largely decorative world map, which merely distinguishes land from water.

Occasionally a complaint seems blatantly contrived, as in 2006, when British newspapers bashed a European Union (EU) initiative to promote cross-boundary cooperation in tourism and environmental protection as "a German-led 'conspiracy of cartographers' [designed] to give Brussels the power to change national boundaries."[4] Politicians wary of surrendering British sovereignty to an international government uncovered an EU Web site on which bland maps with thin borders and nonthreatening pastels highlighted borderland areas eligible for program support.[5] The *Sunday Telegraph* broke the story with a shrill headline warning that a "new EU map makes Kent part of same 'nation' as France."[6] The next day the *Daily Mail* mapped this "Reshaped View of Europe" with bold dashed lines separating five zones outlined in primary colors (fig. 3.1), including bright green for the TransManche (cross-channel) Region, tying Kent to Calais.[7] Although the story faded quickly when readers realized the EU delineations were more than two years old, it underscored the power of maps to exaggerate and inflame.

Perhaps the most telling tale of map-induced hysteria occurred in the early years of World War II, shortly before Germany overran France. The cover of the March 16, 1940, issue of *L'Illustration,* a French weekly news-

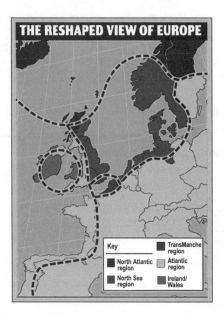

FIGURE 3.1. Map posted on September 4, 2006, on the Web site of the *Daily Mail*, a London tabloid, dramatized a spurious EU threat to British sovereignty.

magazine, pictured two U.S. diplomats, Sumner Welles and Robert Murphy, in the office of French finance minister Paul Reynaud. Resting on an easel near the center of the photo was a large map of Europe (fig. 3.2). More a prop than a focus of discussion, the map was seen by some readers as a scheme to reconfigure Europe's boundaries. German officials in Berlin denounced France for denying the Third Reich's recent acquisitions, and Hitler's allies in Rome were equally incensed that the map ignored Italy's annexation of Trieste after World War I.[8] Equally bewildering but less controversial were missing borders between Denmark and Germany and between Belgium and The Netherlands as well as some odd kinks in the Franco-German frontier—distortions not particularly advantageous for the French.

As amused journalists soon discovered, the apparent conspiracy had more to do with penny-pinching than geopolitical conniving.[9] The map had been printed in 1919, and the French finance ministry used colored chalk to avoid buying a new one—Reynaud had reportedly updated the map himself, with red for Germany's recent conquests and yellow for Russia's. When the red and the yellow turned out solid black on the photo, a censor wary of treating Russia as aligned with Germany ordered the printer

L'ILLUSTRATION

N° 5063 • 16 MARS 1940   PRIX : 5 FRANCS

CONVERSATION DE M. PAUL REYNAUD AVEC M. SUMNER WELLES AU MINISTÈRE DES FINANCES
(au fond, de face, M. Murphy, chargé d'affaires des États-Unis)
Voir les pages 251 et 264.

DANS CE NUMÉRO :

LA FINLANDE DEVANT   LES VISÉES GERMANIQUES SUR LA BELGIQUE
          LES PROPOSITIONS SOVIÉTIQUES   (d'après des documents secrets allemands).

LES GUERRIERS DU MOYEN-ATLAS   UN TORPILLEUR FRANÇAIS ÉPERONNE
          SUR LE FRONT FRANÇAIS   UN SOUS-MARIN ALLEMAND
                                                      (composition d'Albert Sebille).
TROIS NOUVELLES USINES DE GUERRE
          EN FRANCE   LE CONTROLE POSTAL EN MER, ETC.

En hors texte : portrait en couleurs du général PRÉTELAT, par Lucien Jonas.

FIGURE 3.2. Cover of the March 16, 1940, issue of *L'Illustration*.

to make Europe a uniform gray. Thinking that a map without borders just didn't look right, the retoucher added some boundaries, apparently from memory and clearly in haste. The controversy escalated until the French ministry produced the original map and its photographic offspring, which calmed the Germans and the Italians but did little to discredit the widely held belief that maps and photographs don't lie.

More deliberate cartographic erasures can be found on contemporary Arab maps of Palestine that omit Israel as well as on sixteenth- through nineteenth-century European maps of North America that denied the presence of native peoples. Cartographic rhetoric is equally obvious on Argentinean maps that claim sovereignty over the Falkland Islands, a former Spanish colony known in Buenos Aires as the "Islas Malvinas." After separating from Spain in 1816, Argentina intermittently controlled the Malvinas until 1833, when Britain forcibly reasserted an even earlier claim. Maps in Argentina's schoolbooks and on its postage stamps never conceded the loss. As persistent symbols of historical entitlement and national identity, they partly underlie the unsuccessful attempt to retake the islands in 1982.

Lines Matter

Much of an international boundary's power stems from restrictions on cross-border movement and disparities in civil administration and socioeconomic well-being. Occasionally the contrast is apparent from space. While the U.S.-Mexican border clearly separates nations of diverse opportunity, its impression on the landscape is usually less visible than along the boundary between Haiti and the Dominican Republic, which share the Caribbean island of Hispaniola. Both countries are poor, but population pressure and political corruption have been far more catastrophic in Haiti, where rampant deforestation and soil erosion account for marked tonal contrasts on the satellite image in figure 3.3. Because rock and bare soil tend to reflect light, the Haitian side of the border, on the left, bears the imprint of rural overcrowding, imprudent farming on steep slopes, and the destructive harvesting of timber for fuel.[10] In this example an international border amplifies the impact of local property boundaries.

Heavily fortified borders like the Military Demarcation Line (MDL) separating North and South Korea have similarly prominent footprints. Roughly 240 kilometers (150 miles) long, the MDL approximates the positions of opposing troops in July 1953, when the Armistice Agreement was signed. It's also the center of the Korean Demilitarized Zone (DMZ), a swath of land four kilometers (2.5 miles) wide seeded with landmines and rimmed by a 10-foot-tall barbed-wire fence.[11] Reinforcing the DMZ to the south is the Civilian Control Zone, three to twelve miles wide and restricted to troops, heavy weapons, and surveillance systems. North Korea maintains a similar buffer. Despite occasional use of defoliants in years past, the DMZ has become noticeably greener than its surroundings. Iron-

FIGURE 3.3. Black-and-white rendering of a Landsat-5 image captured May 13, 1998, along the border between Haiti (*left*) and the Dominican Republic (*right*).

ically, what's sometimes called the scariest place on earth is now a haven for plants and wildlife endangered by agriculture and urbanization.

Dangerously close to Seoul, a major world city, the Korean DMZ exemplifies the unsettled and often unsafe boundary known as a frontier. Intended as a buffer between combatants, the Korean DMZ is similar to the variable-width Green Line separating Greek and Turkish factions on Cyprus. Comparatively remote frontiers include the "Line of Actual Control" between Chinese and Indian forces in Aksai Chin, a part of Kashmir administered by China but claimed by India, and the Siachen Glacier, 50 miles to the west, where Indian and Pakistani forces have faced off since the mid 1980s.[12] Equally bizarre are the diamond-shaped "Neutral Zone" areas that once separated Saudi Arabia from Kuwait and Iraq. Relics of treaties signed in the 1920s, these largely unpopulated buffer zones were partitioned in 1971 and 1983, respectively.[13] Recent atlases still use dashed lines for Saudi Arabia's southern borders with Yemen, Oman, and the United Arab Emirates because these boundaries have been in dispute or not completely delineated.

While international boundaries marked by fortifications, security fences, or DMZs create distinctive landscapes, borders between low-contrast neighbors are often apparent only in intermittent monuments, warning signs, and official border crossings. An obvious example is the U.S.-Canadian boundary, described pretentiously as "the world's longest undefended border." Policed on both sides by a sprinkling of customs agents and border guards, the boundary is maintained by the International Boundary Commission (IBC), an intergovernmental agency with American and Canadian commissioners and regional offices in both countries.[14] To avoid any misunderstanding about the border's location, the IBC has not only defined the 3,145-mile (5,061-kilometer) land boundary with 5,528 monuments referenced to a common survey system but also connected the intervening line segments with a "vista" of cleared land 20 feet (6 meters) wide. Although chainsaws, bulldozers, and herbicides have replaced handsaws and axes in remote forested areas, a less jarring look is apparent along portions of the vista "groomed" by selective spraying and pruning and landscaped with native grasses and ferns. With similar precision the 2,380-mile (3,830-kilometer) water boundary, mostly through the Great Lakes and along the St. Lawrence River, is described by straight lines between 5,723 unmarked "turning points," anchored to the shore by 2,457 reference monuments. For officials in Washington and Ottawa the border's precise location is not an issue.

Residents of Blaine, Washington, and its next-door neighbor Surrey, British Columbia, are not so sanguine. Government agents wary of smugglers and illegal migrants have seeded the vista with electronic detectors, known to summon a Border Patrol helicopter when nearby homeowners cut their grass. Newcomers are especially vulnerable, as Shirley-Ann and Herbert Leu discovered in 2007, when the IBC ordered the retired couple to tear down their recently built four-foot retaining wall.[15] Although well within their property line, the wall was 30 inches too close to the border. Before spending $15,000 on the project, the Leus had obtained a permit from city officials, who apparently knew nothing of—or had forgotten about—a restriction against building within 10 feet of the border. Nothing in their deed or property survey warned of the ban, and when the IBC was unable to cite a U.S. counterpart to a Canadian law specifically forbidding construction inside the vista, the couple sued, their lawyer broke the story to the media, and President Bush fired the U.S. commissioner, a party loyalist appointed six years earlier.

While embarrassed public officials in Blaine now caution against building too close to the border, don't expect them to issue updated zoning maps advertising the IBC vista. Designed to implement local land-use regulations, municipal zoning maps typically show the international border as nothing more than a bold dashed line along the edge of the map. Community development offices along the U.S.-Mexican border are no less myopic than their northern counterparts, even after the Department of Homeland Security announced plans for a vehicle barrier with multiple layers of security fencing. Although our southern boundary never had a 20-foot-wide vista—large trees and dense shrubbery are rare in the southwest—a buffer of some sort seems likely. One proposal included a 100-yard-wide border zone, wholly on American soil, which would afford access along both sides of the barrier as well as reduce the likelihood of putting it in the wrong place.[16] Mistakes happen, as U.S. Customs and Border Protection, a cog in the Homeland Security bureaucracy, confessed in 2007, after Mexican officials complained that 1.5 miles of a 15-mile concrete-post barrier near Columbus, New Mexico, in 2000 was one to six feet over the line.[17] Apparently the contractor merely followed an existing fence, improperly placed in the late nineteenth century. The estimated cost of tearing out and replacing the encroaching barrier was $3 million—perhaps less if the successful bidder hired undocumented workers.[18]

This mistake might have been avoided had the border with Mexico been inscribed on the landscape as precisely and prominently as the border with Canada. A partial explanation lies in the name of the binational agency responsible for placing and maintaining monuments. The International Boundary and Water Commission is less focused on boundary surveying than on water rights, flood control, water quality, and wastewater treatment, which collectively dominate its staff and budget. Of course, many of its water-related projects are closely related to boundary demarcation insofar as the Rio Grande—the Rio Bravo del Norte on Mexican maps—accounts for 64 percent of the border's 1,954 miles (3,145 kilometers).[19] Free of foliage and much narrower than the Great Lakes, the Rio Grande is a ready-made, self-maintaining boundary vista.

Despite the river's prominence as a landscape feature, it's been an irritatingly ambulatory boundary, subject to catastrophic shifts as well as slow, steady erosion. In 1884 the United States and Mexico signed a treaty that recognized the middle of the river as the international boundary except when "the force of the current" caused "the abandonment of an existing river bed and the opening of a new one."[20] In these instances the original channel bed surveyed in 1852 was to prevail even if it were "wholly dry or . . . obstructed by deposits." Although the treaty reaffirmed the conventional treatment of dynamic boundaries, phrases like "natural causes" and "slow and gradual" invited subjective interpretation as did the river itself by advancing both gradually and abruptly.

In remote, sparsely populated stretches of the border, abandoned meanders create few problems: the border stays where it was, and one country merely occupies both banks of the new channel. But in urban areas like the sister cities of El Paso, Texas, and Ciudad Juárez, Mexico, an ambulatory boundary that's advancing southward progressively as well as in spurts is inherently contentious.[21] In 1895, for instance, Mexico formally claimed approximately 600 acres between the river's positions in 1852 and 1895 in a tract named the Chamizal after a species of desert grass. The United States, which held the tract, attributed the river's advance to gradual erosion, while Mexico cited evidence of avulsion. A further complication arose in 1899, when a diversion channel endorsed by the mayors of both cities as a flood-control measure lopped off a meander that protruded northward. Under terms of the 1884 treaty, the defunct meander bed remained the international boundary while the area within, which became known as Cordova Island, remained attached to Mexico across a narrow passage (fig. 3.4).

Urban growth aggravated the controversy as Cordova Island became an obstacle between the east and west portions of El Paso. In 1911 an arbitration panel consisting of one Mexican, one American, and one Canadian decided that the border west of Cordova Island should follow the river's channel in 1864, located between its positions in 1852 and 1911 (fig. 3.4, *upper*). The United States rejected the commission's decision, and Mexico condemned American arrogance. Even so, the two nations continued to simplify and stabilize other portions of their riverine boundary by exchanging land and constructing concrete channels. The threat of flooding required further adjustment of the channel between El Paso and Juárez, and continued development near the border made American officials wish their government had accepted the 1911 decision.[22]

A compromise was reached in the early 1960s, after President Kennedy sought improved relations with Mexico and engineers concluded that the most efficient position for the river was a new channel through the northern portion of Cordova Island. In an agreement approved in 1963, Mexico gave up 193 acres of undeveloped land in the northern part of the tract, while the United States ceded 264 acres to east and 366 acres to the west

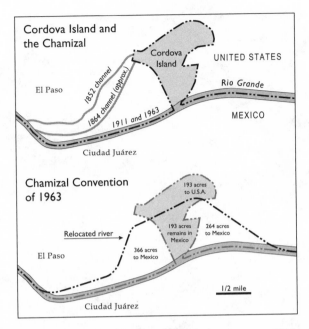

Cordova Island and
the Chamizal

Cordova
Island

UNITED STATES

El Paso

1852 channel

1864 channel (approx.)

1911 and 1963

Rio Grande

MEXICO

Ciudad Juárez

Chamizal Convention
of 1963

193 acres
to U.S.A.

Relocated river

193 acres
remains in
Mexico

264 acres
to Mexico

El Paso

366 acres
to Mexico

1/2 mile

Ciudad Juárez

FIGURE 3.4. Cordova Island and positions of the Rio Grande responsible for the Chamizal dispute (*above*), resolved seven decades later with a concrete channel and related land transfers (*below*).

(fig. 3.4, *lower*).[23] In addition to a more stable international boundary, the new concrete channel was a dramatic gesture of goodwill, for which the federal government paid heavily to buy out property owners and relocate thousands of residents. The map excerpts in figure 3.5 contrast some of El Paso's lost neighborhoods with the spacious sites of a new high school, a new port of entry, and the Chamizal National Memorial, a park and museum commemorating the peaceful settlement of a long-standing dispute. Typical of topographic maps of borderlands, cartographic details stop abruptly at the boundary to acknowledge Mexico's sovereignty as well as lighten the mapmaker's work.

Delimitation, Demarcation, and Defense

Maps can both help and hinder boundary making. Although detailed, reliable cartography is a valuable tool for figuring out where to put a border—diplomats call this delimitation—a simple border considered easy to map

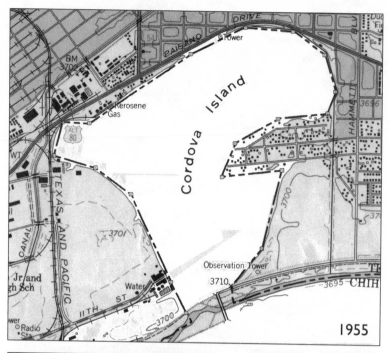

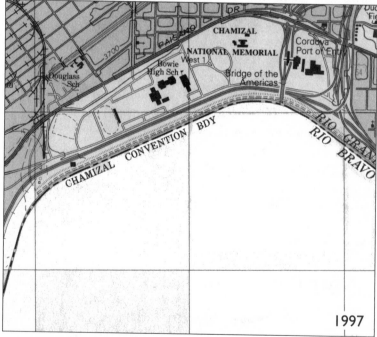

FIGURE 3.5. International border before and after the Chamizal Convention of 1963.

can prove difficult to inscribe on the landscape with surveyor's monuments and nearly impossible to defend. Straight lines are perhaps the most troublesome. Drawn in less than a minute with a straightedge, a direct line between distant end points might entail years of fieldwork if the demarcation team must hack its way through dense forests or repeatedly crisscross rivers and canyons. Particularly problematic is the settle-it-now/map-it-later approach that sidesteps the lack of large-scale topographic maps by tying a boundary to the "source" or "mouth" of a river, the "crest" of a mountain range, or the "parting of waters" between two drainage basins.

Perhaps the most notorious example is the Argentina-Chile border, defined by an 1881 treaty as the line "along the highest crests of [the Cordilleras of the Andes] which may divide the waters, and shall pass between the slopes which descend on either side."[24] This definition worked well along the northern portion of the 3,000-mile boundary but proved contradictory farther south, through Patagonia, which was little explored and poorly mapped. As field parties sent to map the boundary soon discovered, the drainage divide did not always follow the line of high peaks, as the negotiators had assumed. At times the water-parting entered large swamps or lakes well east of the backbone of the Cordilleras, no doubt to the surprise and delight of the Chileans. Equally troubling were substantial areas of interior drainage not tributary to either ocean. Fortunately, the treaty called for arbitration to settle disputes "in which the watershed may not be apparent."

Resolution of the Argentina-Chile boundary dispute owes much to Thomas Hungerford Holdich (1843–1929), a British military surveyor who retired from the Survey of India in 1898, after three decades of service, and was appointed to the arbitration tribunal arranged by England at the request of the two countries. As an advocate for taking advantage of whatever geographic barriers Mother Nature might offer, Holdich believed that disputes arising from vague terminology are best resolved by a neutral demarcation commission authorized to make adjustments in the field and redraw the map as needed. He favored lines that were clearly visible, easy to mark, and difficult to cross. For helping settle numerous boundary disputes in Asia and Latin America, Holdich was awarded a knighthood and elected president of the Royal Geographical Society. His book *Political Frontiers and Boundary Making*, published in 1916, anticipated the postwar reconfiguration of Central Europe and the Middle East. In its preface, he chided academic contemporaries for advocating boundaries based on race and culture, rather than a "practical delimitation [designed] to limit unauthorized expansion and trespass."[25]

Holdich's preference for defensible barriers reflects the importance of fortifications in the largely land-based warfare of the early twentieth century as well as a wariness of the lust for *lebensraum* (living space), the central concept behind Adolf Hitler's territorial aggression in the 1930s. "Personal ambition backed by military strength and the possession ... of the greatest military efficiency has led to a vast multitude of wars in the past," Sir Thomas warned, and even where "the expansion of a people has been the result of a too crowded increase of the population creating a necessity for new fields of national development, it is always the military capacity of the people which decided the success of the movement."[26] National defense demanded a strong army, and strong borders made the soldiers' work much simpler: "boundaries must be barriers—if not geographical and natural, then they must be artificial, and [as] strong as military device can make them."[27]

In distinguishing between natural and artificial boundaries, Holdich noted that rivers and drainage divides, though lacking the strategic advantage of a mountain range with a clearly marked ridge crest and few passes, at least minimized conflict by simplifying demarcation. Truly artificial boundaries, he argued, "are the most expensive to demarcate [and the] most difficult to maintain, [and] no diplomatist should ever agree to artificial demarcation until he is satisfied by trustworthy evidence in the shape of accurate local surveys that a natural boundary is impossible."[28] Detailed local maps are essential in settling a boundary dispute, he maintained, because "a boundary is but an artificial impress on the surface of the land, as much as a road or a railway, and, like the road or the railway, it must adapt itself to the topographical conditions of the country it traverses."[29]

While terrain makes some boundaries arguably more natural than others, few are as blatantly artificial as the Great Wall of China, 4,000 miles (6,400 kilometers) long and built in several sections, starting in the fifth century BC. Used to defend China's northern border for more than 2,000 years, it was breached in 1644 after Manchu invaders bribed a military commander.[30] It's now a must-see tourist attraction. Less prominent is Hadrian's Wall, a 73-mile (118-kilometer) barrier built by the Romans across northern England in the second century AD.[31] Abandoned shortly after the death of Emperor Hadrian, whose successor built a shorter, less durable wall farther north, it too survives as a tourist site. A much younger barrier is the Maginot Line; less a wall than a line of fortifications, it kept the Germans from invading France directly, across their common boundary, but did nothing to thwart the Wehrmacht's end run through Belgium. By contrast, the Berlin Wall, a true wall 28 miles (45 kilometers) long, is famed for keeping people in rather than out. Never totally effective in deterring escape to West Berlin, it was an icon of Soviet heavy-handedness during the Cold War and a symbol of Western triumph after its removal in 1989.

Impenetrable barriers crafted with masonry or barbed wire are still surprisingly common—a sidebar in the March 2005 issue of *Atlantic Monthly* lists 10 of them, including the Korean DMZ and "a system of high-tech barriers" along 70 miles of the U.S.-Mexico border.[32] None are as controversial as the Security Fence initiated by Israel in 2002 after an epidemic of suicide bombings.[33] Intended as a barrier between Israel and the West Bank, the fence darts eastward at several points to incorporate Jewish settlements established in the decades following the Six Day War in 1967 (fig. 3.6). The resulting "Seam Zone" between the fence and the former Israel-Jordan boundary, established after the first Arab-Israeli War and known as the Green Line, severely hampers travel between home and work for tens of thousands of Palestinians. Fearing a permanent loss of territory, they appealed to the International Court of Justice in The Hague, which in 2004 declared the fence a violation of international law and called for its removal. Israel refused to accept the ruling and chided the court for ignoring the barrier's effectiveness in reducing terrorist attacks on civilians. Although the Ministry of Defense revised the route several times as a humanitarian gesture, right-wing politicians determined to protect the settlements undermine the government's contention that the fence is only temporary.

Maps of the West Bank reveal a jumble of contested boundaries as well as interspersed Hebrew and Arabic place names, sometimes one of each for the same city. Jerusalem, for instance, is parenthetically Al Quds, while Nablus (Arabic) is also Shekhem (Hebrew). Central Intelligence Agency maps describe the region between the Green line and the River Jordan as "Israeli-occupied—status to be determined." Curious features include several patches of "No Man's Land," which once served as a buffer between Israeli and Arab forces (fig. 3.7). Although Israel annexed these areas after the Six Day War and doesn't consider them part of the West Bank, CIA maps portray them with a light purple tint as wide parts of an otherwise

FIGURE 3.6. Encroachments of the Israeli Security Fence beyond the Green Line into the West Bank. Because the route of the fence was adjusted several times between 2005 and 2007, the "Seam Zone" shown is a generalized representation of the threat perceived by Palestinian Arabs.

thin border. Particularly troublesome as obstacles to peace are numerous bright blue zones identified in the map key as "Israeli-developed area (civilian) beyond the 1949 armistice line." Western diplomats consider peace unlikely unless Israel withdraws its troops and settlers and the Palestinians establish a stable, independent nation—which means that Israel must somehow placate religious ideologues convinced that the West Bank is rightfully theirs while its Arab neighbors shun the infighting and extremism that undermined Yasir Arafat's Palestinian Authority. If an agreed boundary ever separates Israel and Palestine, its demarcation will no doubt include, for a while at least, a visible barrier like the Security Fence.

Despite cartography's obvious role in settling territorial disputes, maps of past boundaries and obsolete place names are tempting fodder for anyone eager to resurrect old land claims or demand reparations. Particularly worrisome are Israeli maps stamped with contrived Hebrew names

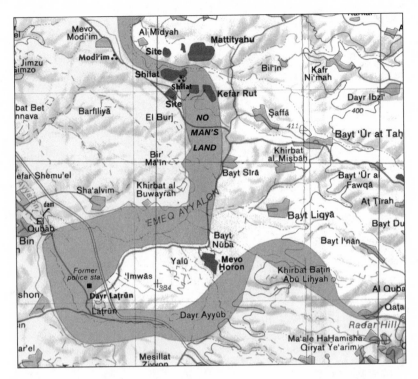

FIGURE 3.7. Portion of the boundary between Israel and the West Bank roughly 10 miles west of Jerusalem widens into "No Man's Land." Area shown is approximately 11 miles left to right. Israel is on the left, and the West Bank is on the right. The dark gray patches are Israeli settlements, as shown on the 1983 map. A 1992 edition shows several more settlements within this rectangle.

for hills, streams, and other geographic features and Palestinian "counter maps" circulating in cyberspace to perpetuate hamlets and olive groves obliterated by Israeli bulldozers.[34] Because cartographic evidence looks convincing, intentionally biased maps lurking in attics or archives can revive ethnic rivalry decades after a conflict is resolved.

Cartographic boundaries make similarly persuasive arguments at the broader scale of regions and continents. In the nineteenth century, for instance, European powers intent on partitioning Africa and colonizing parts of Asia would declare a sphere of influence. As Sir Thomas Holdich observed, "When any nation announces of any extent of territory 'this is within my sphere of influence,' it practically sets up a warning to trespassers much in the same way that a Scottish landowner warns casual pedestrians off his grouse moors."[35] Like the sales territories laid out to avoid turf wars among sales reps, spheres of influence let the colonizers focus on exploitation.

A related concept is the panregion formed when nations on one continent acquire political and economic influence on an adjoining continent. The German geopolitical theorists who developed the notion in the 1930s saw Eurafrica as a panregion ripe for German hegemony.[36] Nazi propagandists extended the concept in *Facts in Review*, a magazine published in New York between 1939 and June 1941 by the German Library of Information. In April 1941, as the United States drifted toward entering World War II on the side of Britain, the magazine ran a small map delimiting panregions focused on the United States, Europe, Russia, and Japan (fig. 3.8). The bold line encircling North and South America was a subtle reminder of President James Monroe's 1823 warning to European powers: butt out of the Americas, or at least don't establish any new colonies here. Not so subtle was the suggestion that the United States had no business interfering with the Axis powers in Eurafrica and the Pacific. A plausible argument perhaps, but a bit late and hardly influential.

Pockets of Sovereignty

How big must a sovereign territory be to have its boundaries mapped? The answer depends at least partly on the scale of the map and the size of the country. Like other tiny nations, Andorra (174 square miles) doesn't show up on page-size world maps, which barely have room for Belgium, yet its boundaries appear on maps of France or Spain. Similarly, Vatican City (109 acres), which at best warrants a point symbol and label on a map of Italy, acquires an international border with a distinctive shape on a map of Rome. The boundaries of Andorra and Vatican City are symbolized as international borders because other independent countries acknowledge their sovereignty, which is especially important for landlocked "microstates."[37] The Vatican exchanges diplomats with Italy, the United States, and many other countries, but it's not a member of the United Nations, whose smallest constituent is now Nauru (9 square miles), a Pacific island near the equator about 15 degrees east of New Guinea. This distinction once belonged to San Marino (23 square miles), which is wholly surrounded by Italy and a UN member since 1992.

A step down from internationally recognized sovereignty is the extra-

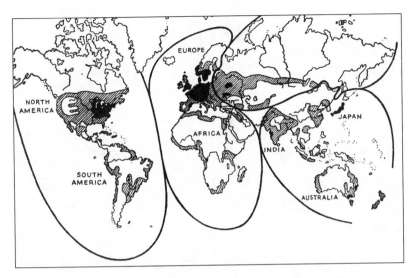

FIGURE 3.8. Bold lines separating "spheres of influence" advise the United States to stay out of World War II.

territorially enjoyed by another Italian enclave, the Sovereign Military Order of Malta (SMOM), whose leaders fled Malta when Napoleon invaded in 1798. Its territory now consists of two small properties in Rome—the Palazzo Malta in Via dei Condotti 68 and the Villa Malta on the Aventine.[38] Although extraterritoriality means little more than the freedom from local authority accorded embassies, consulates, and military bases, the SMOM issues passports, prints postage stamps and mints coins (primarily for collectors), and flies its own flag.[39] Part religious order and part government in exile, this wannabe country sends an "observer" to the UN, but its lot lines, like embassy perimeters, are not mapped as international borders.

Size and sovereignty affect the cartographic treatment of the Oneida Nation Territory, an Indian reservation in Madison County, New York, about 30 miles east of where I live. Its occupants have a distinguished but tragic history. Their ancestors were the only one of the Six Nations of the Iroquois to support the colonists in the American Revolution. The war devastated the Oneidas' homeland, and although the new republic guaranteed them a reservation of 6 million acres, by 1788 pressure from New York officials to sell land to the state had reduced their holdings to 300,000 acres.[40] The Indian Nonintercourse Act of 1790 prohibited the states from buying native lands without federal approval, but New York paid no heed

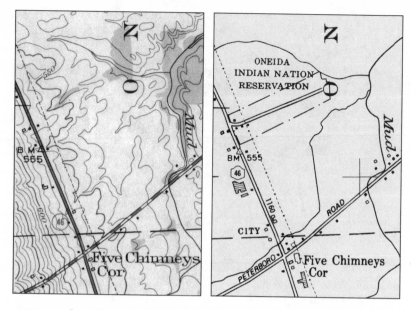

FIGURE 3.9. Ignored on a 1955 federal topographic map (*left*), the 32-acre Oneida reservation gained cartographic recognition on a 1978 New York State quadrangle map (*right*).

and continued to acquire tribal land. Many Oneidas decamped for Canada and Wisconsin, and tribal holdings in the state declined to 4,509 acres by 1839, 933 acres by 1843, 350 acres by 1890, and a mere 32 acres by 1920.[41] Although the New York group remained an officially recognized "sovereign Indian nation," the first U.S. Geological Survey topographic quadrangle map of the area, published in 1895 at 1:62,500,[42] ignored the Oneidas' 32 acres as did the most recent revision, issued in 1955 at 1:24,000 (fig. 3.9, *left*). But 23 years later, the New York Department of Transportation, which had started its own statewide series of quadrangle maps, included the "Oneida Indian Nation Reservation" on a map at the same scale (fig. 3.9, *right*). Although most reservations are far larger, a 32-acre tract is not too small for a boundary line on a large-scale map.

Size more than sovereignty explains the territory's earlier omission. Federal topographic maps have long recognized the importance of tribal lands: a dot-and-long-dash "reservation line" symbol identical to the boundary line on the New York map (fig. 3.9, *right*) is among "standard symbols" explained on the back of late nineteenth-century U.S. Geological Survey (USGS) quadrangle maps, and an identical symbol is among the "Boundaries, Marks, and Monuments" illustrated in the Geological Survey's 432-page *Topographic Instructions,* published in 1928.[43] Broadly intended for "national or State parks, forests, monuments, bird and game preserves, Indian, military, or lighthouse" lands,[44] the symbol delineated the 7,300-acre "Onondaga Indian Reservation," south of Syracuse, on a 1900 USGS map.[45] Unwilling to rely solely on its field topographers, the Geological Survey required that "Indian reservations should be verified by the State maps of the General Land Office and as to their boundaries by existing maps, by the descriptions given in the acts creating them, by reference to the township plats of the General Land Office, or through inquiry at the Office of Indian Affairs."[46] Even so, if federal mapmakers didn't know about the tiny tract, they had no reason to seek verification.

That the 1978 New York map acknowledged the 32-acre reservation was not a matter of better research or sharper observation. By the 1970s state officials and the local news media were well aware of the Oneidas, who had sued to reclaim 300,000 acres conveyed to the state under an illegal 1795 treaty. Because New York had not obtained federal permission, the tribe's attorneys argued, that transaction was void and the land still belonged to the Oneidas. Their first significant victory came in 1974, when the U.S. Supreme Court declared the case within the jurisdiction of the federal courts.[47] Eleven years later the high court endorsed a lower-court ruling that the conveyance was illegal and the Oneidas deserved compensation.[48]

Although the details of compensation were still in litigation, the Oneida Indian Nation used its newfound leverage to establish the Turning Stone Resort and Casino six miles north of the original reservation, just off Exit 33 on the New York Thruway. Profits from the casino, a bingo parlor opened in 1979, and a local chain of convenience stores selling tax-free gasoline and cigarettes supported an expanded array of services to tribal members as well as the piecemeal purchase of land within the boundary of the original 300,000-acre reservation. By 2005, the nation had acquired 17,000 acres in Madison and Oneida Counties—less than 1.5 percent of the two-county area but a potential threat to the tax base in the minds of landowners who feared substantially higher tax rates. Sovereign Indian territory is exempt from local property taxes, and local officials were concerned that a substantial damage settlement would support a much larger buyback. What's more, the Oneidas were not collecting sales tax, a portion of which was earmarked for local government. Although the

Oneida Nation was now a major employer of non-Indians and contributed significantly to local school districts, the counties and a few municipalities continued to challenge the Oneida's victory.

In 2005 the Supreme Court weighed in again, this time to "preclude the Tribe from unilaterally reviving its ancient sovereignty."[49] The Oneidas had waited too long to seek relief, eight of nine justices concluded, and state and local authority was too well established to be undermined by a fragmented expanded reservation. The nation was still entitled to the monetary compensation affirmed by the 1985 decision, but its rights as a landowner were little different from those of any other resident—with one exception. By putting their holdings into a federal land trust, the Oneidas could exchange their right to sell for freedom from local property taxes. The Bureau of Indian Affairs, which would administer the trust, commissioned a "Draft Environmental Impact Statement," which documented the fragmented character of the nation's 17,000 acres, outlined various options, and expressed concern that additional purchases "be reasonably contiguous to current property groupings, thereby minimizing jurisdictional issues between Nation and non-Nation land."[50] As I write, the case remains in litigation. Although less sovereign then the limited autonomy allowed officially recognized Native American tribes, a large Indian land trust in the Oneidas' name would surely merit cartographic recognition.

....

This chapter's examination of the delineation, demarcation, and impact of international borders underscores the boundary map's symbiotic tie to sovereignty, a fiercely territorial concept that depends on fencing out other countries' armies, tax collectors, and emigrants. The next chapter extends this exploration with a closer look at boundaries imposed by imperial cartographers and peace conferences.

CHAPTER

FOUR

....

ABSENTEE
LANDLORDS

Territorial boundaries imposed by colonial powers or peace conferences are often less harmonious than borders allowed to evolve between quarrelsome neighbors. Colonizers eager to get on with exploitation typically tie their delineations to astronomy, geometry, or physical geography, each a name brand of scientific objectivity, while peacemakers anxious to return to lucrative trading prefer lines of separation based on ethnicity, language, race, or religion, all widely assumed to confer geopolitical stability. This combination of convenience and naïveté accounts for artificial borders with notoriously tragic outcomes in Africa, the Balkans, and the Middle East. As this chapter shows, the seductiveness of artificial boundaries stems from the simplicity of cartographic illustration and a naive faith in drawing a boundary now and sorting out its details later.

Carving Out Colonies

Explorers who plant flags on newfound beaches fulfill the first of two traditional requirements for claiming overseas territory: discovery and occupation. Although published sightings were generally acknowledged by colonial competitors, who invoked the discovery principle themselves, the rules of the game required occupation of some sort, with forts, trading posts, and permanent settlers as well as treaties with naive native inhabitants, too readily bribed or coerced into becoming subjects or junior partners of a distant sovereign. Firearms gave Europe's colonial powers a decisive edge, and if armed slaughter were distasteful, imported epidemics could always stifle resistance. No point, though, in killing them all off insofar as indigenous inhabitants experienced in battling their neighbors could help repel incursions from another country's colonial army and its own native allies. Because war often proved costly, if not counterproductive, colonizers quickly learned the value of the map-based treaty as a license to occupy and exploit.

The earliest cartographic accord affecting the Americas was concluded by Spain and Portugal in 1494, two years after Christopher Columbus landed in Caribbean and claimed the "New World" for his Spanish sponsors, Ferdinand and Isabella. Named after a city in central Spain, the Treaty of Tordesillas followed Pope Alexander VI's June 1493 pronouncement dividing all new and yet-to-be discovered non-Christian territory between the Portuguese and the Spaniards. A north-south line of demarcation running from pole to pole was anchored to a point 100 leagues west of the Azores, a Portuguese possession once thought to include the westernmost island in the Atlantic. Separating Portuguese territory to the east from Spanish lands to the west, the line was intended to split the distance between the Azores and Columbus's recent discovery. Portugal's King John II welcomed the ecclesiastical franchise but feared the papal meridian was too far east—or as some scholars suggest, not far enough west to keep Spain away from Africa.[1] His negotiators persuaded their poorly prepared Spanish counterparts to move the boundary to a point 370 leagues west of the Cape Verde Islands. Although other European sovereigns never accepted the treaty, Portugal benefited significantly six years later when Pedro Álvares Cabral, sent to explore waters east of the new line, stumbled upon Brazil.

Map historians recognize the Tordesillas Line as a milestone of renaissance cartography. In addition to signaling an emerging awareness of territorial sovereignty, it helped jumpstart official government cartography in Portugal and Spain, each with its own strategy for drawing the boundary. Like many lines of anticipation, the agreed meridian was expediently ambiguous, its exact location to be fixed by a joint commission that never met.[2] And because sixteenth-century explorers could not reliably estimate longitude or measure distance at sea, cartographers lacked the data needed to relate the line's position to their slowly evolving, wildly disparate images of islands and coastlines on the far side of the Atlantic. That didn't stop Italian diplomat Alberto Cantino, sent to Lisbon to suss out Portuguese triumphs, from plotting the line on a 1502 map touting Portugal's claims to both Brazil and Newfoundland, named by John Cabot in 1497.[3] Smuggled back to Italy, the Cantino chart was appreciated for its impressively accurate depiction of the African coast, along the route to India pioneered by Portuguese navigators.

Respect for the treaty boundary depended on one's religion. Non-Catholic Britain, for instance, had no qualms in claiming Newfoundland as its first North American colony in 1583. By contrast, Portuguese and Spanish cartographers enthusiastically continued the Tordesillas Line northward over the pole and then southward across the Pacific. The resulting antimeridian stifled Spanish interference with Portugal's seizure of Malacca, on the Malay Peninsula, in 1511, and its subsequent colonization of the nearby Molucca Islands, also known as the Spice Islands—Columbus's intended destination and now part of Indonesia. A grossly inaccurate estimate of longitude situated the islands 17 degrees west of the antimeridian and well inside Portuguese territory. A century of further exploration yielded a more accurate world map that put the colony right on the borderline.[4]

That Spain could claim at least nominal sovereignty over much of its vast papal land grant was an impressive feat, captured cartographically by its official historian, Antonio de Herrera y Tordesillas. (His name reflects paternal ties to the treaty city.) Published between 1601 and 1615, Herrera's detailed four-volume history of the Spanish colonies includes a small foldout map (fig. 4.1) on which the Tordesillas Line and its counterpart 180 degrees to the west bracket the Americas (except for Brazil) and much of the known Pacific. Simple maps like these helped Europe's educated elite visualize the possibilities and privileges of empire.

An ever more aggressive cartography advertised territorial claims. Powerful navies and persuasive diplomats were essential, to be sure, but an overseas empire depended on authoritative descriptions of occupancy and ownership asserted by boundary lines and place names and reinforced by detailed coastlines, realistic terrain, and geometrically precise grids. Like Madison Avenue hucksters, cartographic propagandists occasionally suc-

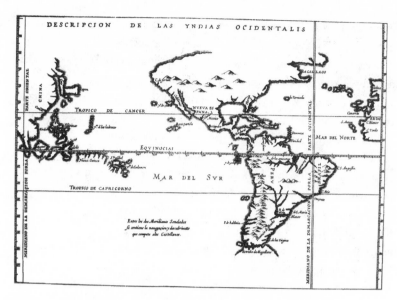

FIGURE 4.1. Descripción de las Yndias Ocidentalis, printed in 1622 from a copperplate engraving based on an earlier map, engraved in wood for Antonio de Herrera y Tordesillas's *Historia general de los hechos de los Castellanos en las islas y tierra firme del Mar Oceano.*

cumbed to blatant exaggeration. Few were as outrageous as John Huske, who extended British colonial boundaries westward across what was widely recognized as French territory—assuming, of course (as colonizers usually do), that native territories didn't count. Included in Huske's 1755 book *The Present State of North America,* the illustration (fig. 4.2) was pompously titled "A new and accurate map of North America: wherein the errors of all preceding British, French and Dutch maps, respecting the rights of Great Britain, France & Spain & the limits of each of his majesty's provinces, are corrected." French forts, which could not be ignored, were surrounded by dotted lines and treated as encroachments.[5]

Until the early nineteenth century Europe's seafaring nations viewed Africa largely as an obstacle on the way to greater wealth in India, the East Indies, China, and Japan. Hazards and landmarks along the coast were charted, of course, and ports along the Atlantic coast supplied slaves for plantations in North and South America as well as food and precious metals for markets in Europe.

Recognition of Africa's potential for plantation agriculture and mechanized mining inspired sporadic efforts to explore the interior, establish

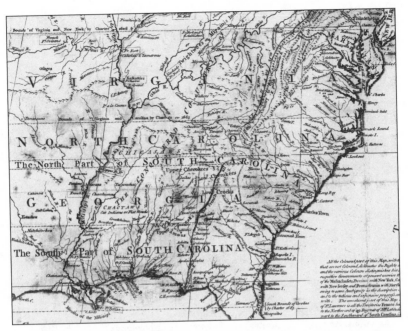

FIGURE 4.2. Excerpt from John Huske's "A new and accurate map of North America," included with his *Present State of North America* (London: R. & J. Dodsley, 1755).

trading posts and forts, and suppress native resistance through treaty or military force. Navigable rivers linking inland areas with oceanic ports were a windfall for European traders, whose successes inspired an influx of civil administrators, soldiers, missionaries, and eventually surveyors. Even so, imperialism was comparatively informal through the 1870s, when a newly unified and assertive Germany sought a share of the continent's wealth.[6] To circumvent growing tensions in the borderlands between competing spheres of influence, German chancellor Otto von Bismarck convened the Berlin Conference of 1884. Inaccurately credited with partitioning Africa, the conference precipitated the fabled "Scramble for Africa" by formalizing relationships among European powers, who agreed that colonial claims were groundless unless backed up by effective occupation. By 1914, the Berlin ground rules had enabled seven countries (Belgium, Britain, France, Italy, Germany, Portugal, and Spain) to own and occupy all but two African territories, Ethiopia and Liberia.

Effective political control demanded distinct, mutually accepted borders, first delimited on paper by diplomats and then demarcated on the ground

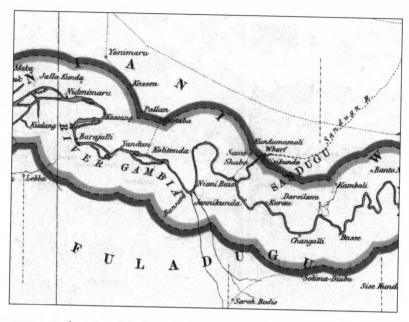

FIGURE 4.3. A portion of Gambia, a long, narrow British colony flanking the River Gambia. Area shown is approximately 70 miles from west to east.

by international survey teams. A boundary commission's task was comparatively easy when suitable maps of the interior allowed an unambiguous anchoring of boundary lines to navigable (and preferably nonambulatory) waterways. "Lines of water parting" between clearly defined rivers could also work, except in flat terrain, where drainage divides are notoriously ambiguous. And as demonstrated by a map of Gambia (fig. 4.3), established in 1888 as a British colony surrounded by French territory, a boundary could even be demarcated a fixed distance from a river bank—10 kilometers in this case—as long as surveyors committed to pragmatism as well as precision had the final say. In Gambia's case, the British and French agreed to mark key points along the river, draw circles with a 10-kilometer radius around these markers, and approximate the intersecting arcs by a series of chords (straight lines between two points on the same arc).[7]

More problematic were straight-line boundaries anchored to a particular meridian or parallel or tied to a pair of widely separated end points. Astronomical and geometric boundaries were especially troublesome in rugged, heavily vegetated terrain with significant gravitational and magnetic anomalies. "The fatal facility with which such lines are ruled upon

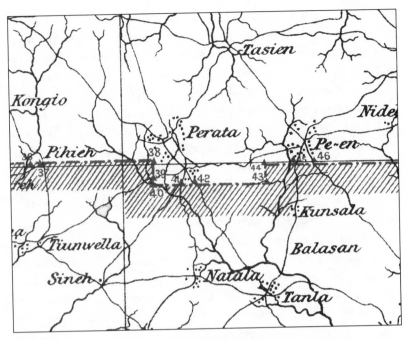

FIGURE 4.4. Adjustment of the boundary between the Gold Coast and French Soudan situated the village of Perata entirely within French territory. Area shown is approximately 21 miles from west to east. Numbers represent beacons placed to mark the boundary.

a map and embodied in a treaty" was particularly frustrating for Edmund Hills, a major in the Geographical Section of Great Britain's General Staff.[8] In a 1906 address to geographers at Cambridge University, Hills railed against "artificial boundaries" conceived by "diplomatists, whose knowledge of geography may be small and whose knowledge of practical survey work is *nil.*"

Hills was also wary of boundaries through inhabited areas and recommended, "wherever possible, respect[ing] existing ethnological or tribal boundaries."[9] Deference to local settlements made sense—why antagonize residents, who could easily sabotage the surveyors' carefully placed and pointedly visible beacons? Whenever a chord along the border between Gambia and Senegal intersected a village, the official boundary was moved 100 meters beyond the outer edge of settlement, away from the majority of villagers, and new buildings were banned within 100 meters of the line.[10] Figure 4.4 describes a similar accommodation by the British-

French boundary commission that refined the border between Britain's Gold Coast colony (now Ghana) and French Soudan. Because the village of Perata was largely north of the negotiated boundary along the parallel at 11° N, the demarcated boundary was placed one kilometer to the south and then continued eastward six kilometers before rejoining the 11th parallel. By contrast, the village of Pe-en required no adjustment.

Despite reasonable accommodations at the local level, the European partition of Africa left a legacy of division and conflict. Although the wave of independence that swept the continent in the decades following World War II radically altered the political map's names and colors—neighboring nations previously part of the British Empire could not both remain red—its national boundaries were inherited from the possessions, protectorates, and provinces of European empires. Understandably, a new name, democratic elections (or their pretense), and a seat at the UN could not unify dissimilar peoples lumped together by artificial boundaries. Civil wars, famine, and mass migrations followed independence in Angola, the Democratic Republic of Congo, and other countries where tribes with diverse traditions and ancient grievances were no longer kept in check by colonial governors tutored in administrative manipulation and backed up by the mother country's military.[11] Sadly, reconfiguration of postcolonial boundaries along tribal lines is no longer an option: sovereign states don't like to give up territory or split off new nations, and Europe's colonizers undermined unification by imposing their own languages on coherent groups like the Yoruba, divided now into English- and French-speakers by the colonial boundary between Nigeria and Dahomey (now Benin). According to George Demko, formerly the chief geographer at the U.S. Department of State, "the main cause of conflict is no longer boundaries—it's corruption, lack of education, and lack of enlightened political leadership and national cohesion."[12]

Partitioning for Peace

If the peace conference that followed World War I reliably reflects expert efforts to reconfigure boundaries, it's unlikely that a massive humanitarian intervention could successfully redraw the map of Africa. While the delegates who gathered in Paris in 1919 rightly blamed dysfunctional borders for some of the tensions that precipitated the war, their efforts to restructure national territory in Europe, Africa, and the Middle East were weakened by conflicting agendas, competing theories of boundary making, and an understandable reluctance to restructure regional demography—a questionable tactic whether you call it population transfer, soft partition, or ethnic cleansing.

A key player at the Paris Peace Conference was Woodrow Wilson, the American president whose resume included a Ph.D. from Johns Hopkins, the presidency of Princeton University, and the governorship of New Jersey. A political scientist by training, Wilson favored letting geographically coherent ethnic groups choose between becoming independent or remaining part of a larger multiethnic nation. Confident of an Allied victory, he wanted the American delegation to be well informed. In August 1917, four months after the United States entered the war, Wilson called for a confidential group of academic experts to collect data, prepare briefings, and draft base maps.[13] Dubbed "The Inquiry" (purposely less conspicuous than its official name, "The Commission of Inquiry"), the hush-hush endeavor was led by Isaiah Bowman, director of the American Geographical Society, which became the group's headquarters.[14] In addition to preparing numerous reports on the war zone's history, geography, and ethnic makeup, the Inquiry's 150 collaborators assembled a huge collection of maps, which accompanied the president across the Atlantic in December 1918.

So many maps were needed because the defeated Central Powers (Germany, Austria-Hungary, and the Ottoman Empire) had territory on three continents. While the British and French delegates took the lead in divvying up Germany's colonies in Africa and in setting up their own so-called mandates (plus a few independent Arab states) in the Middle East, the Americans focused on Europe, particularly Eastern Europe and the Balkans, where dozens of national groups clamored for independence or more land. Bowman's book *The New World: Problems in Political Geography*, published in 1922, includes a map (fig. 4.5) enumerating 32 separate "overlapping territorial claims," about two-thirds of which involved the victorious Allies and their former enemies.[15] The remainder involved only the Allies or "neutrals," such as Russia and Poland. Swept into the conflict, these volatile areas were a logical part of any lasting peace. Because the petitioners at Paris exaggerated their territorial entitlements, Bowman prudently mapped their claims "not in their most extreme but in their most conservative forms."

Drawing boundary lines that respect ethnic traditions, including lan-

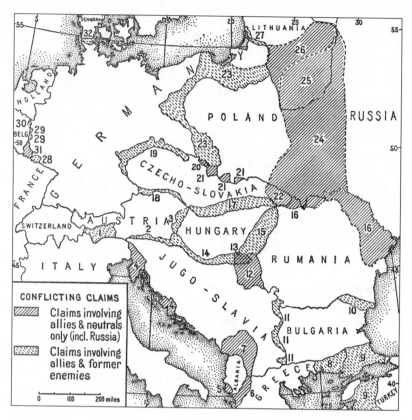

FIGURE 4.5. Isaiah Bowman counted 32 separate overlapping territorial claims in Central Europe.

guage and religion, is famously difficult when dissimilar groups are interspersed. Cultural geographer Ellen Churchill Semple described the problem in a 1907 map (fig. 4.6) of ethnic regions in Central Europe. Most conspicuous are the German enclaves, which extend eastward across Poland into Russia, northward into Russia's Baltic provinces, and southward into the lands of the Czechs, Slovaks, and Magyars (Hungarians). Although German stock seemed least likely to remain within a geographically concise homeland, the map also shows Polish enclaves among the Lithuanians and Germans, and Slovak enclaves among the Magyars. A more detailed breakdown, best mapped at a larger scale able to capture differences among urban neighborhoods, would confirm an even more pervasive intermingling of peoples dissimilar in speech, physical appearance, and customs. Equally troublesome to political geographers was the

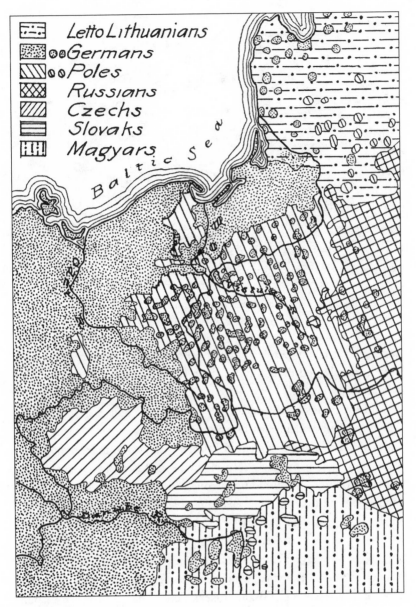

Legend:

- Letto Lithuanians
- Germans
- Poles
- Russians
- Czechs
- Slovaks
- Magyars

Baltic Sea

FIGURE 4.6. Pre–World War I ethnic boundaries in Central Europe.

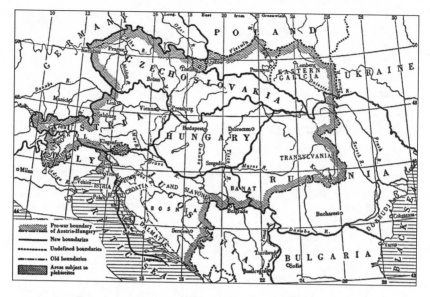

FIGURE 4.7. Disposition of the former Austro-Hungarian Empire according to a map published in 1921 by Charles Seymour, a member of The Inquiry.

apparent mismatch between ethnicity and sovereign territory. As Semple noted, "Rarely, and then only for short stretches, do political and ethnic boundaries coincide."[16]

Did the Paris conference actually improve the odds for world peace? Maybe not, suggested Isaiah Bowman, who observed that replacing 8,000 miles of boundary line in Central Europe with 10,000 miles of boundary line—3,000 of them newly reconfigured—had actually lengthened the zones of friction. "Every additional mile of new boundary," he mused, "has increased for a time the sources of possible trouble between unlike, and in the main, unfriendly peoples."[17] In addition to lengthening the zones of latent conflict, partition of the former Austro-Hungarian Empire (fig. 4.7) created several potentially troublesome polyglot states. In addition to a much expanded Romania, which acquired Transylvania (best known to most of us as the home of Bram Stoker's Count Dracula), the new map included Czechoslovakia and Jugoslavia (later Yugoslavia), whose two-part names forecast their disintegration in the 1990s.

Czechoslovakia was an enigma. Although Czech and Slovak are similar languages, the Czechs were more advanced than the Slovaks, with whom they shared mistrust of the Hapsburg monarchy and an interest in industrial capitalism. Although the new country's name suggests an amalgam of just two peoples, a third of its 14 million inhabitants were neither Czech nor Slovak. Strict adherence to Wilson's doctrine of self-determination would have given Hungary the southern part of Czechoslovakia, where most residents were Magyars, but a long border along the highly navigable Danube provided an important link between the new country's eastern and western components as well as external markets for its growing industries.[18] Self-determination was only one of Wilson's famous Fourteen Points, which also favored economic autonomy.

Jugoslavia seems an even less likely amalgam than Czechoslovakia. Although the defeat of Austria-Hungary had inspired Pan-Slavic nationalists to cobble together the Kingdom of Serbs, Croats, and Slovenes and proclaim its independence in December 1918 under the Serbian monarch King Peter I, the Paris delegates might well have delineated separate national territories for the region's ethnic factions. That the peace conference approved a multiethnic Jugoslavia reflects the active lobbying of Serbian geographer Jovan Cvijić, sent to Paris as an official delegate.[19] Cvijić was well known among the members of the Inquiry: in 1918, the *Geographical Review*, which Bowman edited, published his argument that the Serbs, the Croats, the Slovenes, and the Montenegrins—collectively the Jugo-Slavs (meaning southern Slavs)—were a coherent ethnoracial group. A large foldout map, on which contrasting colors differentiated "zones of civilization," reinforced his point that the Jugo-Slavs had "left a deeper impress than others" on the Balkan Peninsula.[20] Endorsement of Cvijić's reasoning exposed self-determination as a paradoxically flexible concept, able to accommodate both macro and micro notions of "self."

Yugoslavia, which began to disintegrate soon after the death of Marshal Tito in 1980, is the poster child for unstable multiethnic countries. Its legacy includes tens of thousands of Yugoslav citizens who perished in the name of "ethnic cleansing," thousands of additional casualties during the "peacekeeping" intervention of the 1990s, and a half dozen newly sovereign fragments: Slovenia, Serbia, Croatia, Macedonia, Montenegro, and the ominously named Bosnia and Herzegovina. Added to the misery are hundreds of thousands of refuges, whose relocation, however coerced, contributed to greater homogeneity, if not greater stability, at both the place they left and the place they settled.

Lumping together antagonistic peoples invites violent instability. That was a key concern of policy analysts who advocated "soft partition" as a solution to civil war in Iraq, where the American-led removal of Saddam

Hussein unleashed latent hostilities among the country's Sunni Arabs, Shiite Arabs, and Kurds.[21] In contrast to a "hard partition" of Iraq into three independent (and still mutually hostile) neighbors, a soft partition would create three semiautonomous and largely homogeneous regions, each responsible for its own security. Misleadingly simplistic at first glance, soft partition carries numerous caveats and conditions, including just compensation for persons relocated, fair sharing of oil revenues among regions, and an efficient system of checkpoints and identity cards for controlling movement between regions. When a boundary must serve as a fence, where to put the gates is no less important than where to draw the line.

Cutting Up Antarctica

Perhaps the most revealing cartographic claims to sovereign territory center on Antarctica, the last continent to be explored and the last to be occupied year-round, if only by scientific observers. Although Gerard Mercator's 1569 opus and other sixteenth-century world maps include a fictitious southern continent—a lucky guess attributed to Ferdinand Magellan—the Antarctic mainland was not sighted until around 1820 by a Russian, a Briton, or an American, depending on whose historians you choose to believe. Accurate maps of the coastline emerged in the late nineteenth century after the harpoon gun expanded whaling in the Southern Ocean, but Antarctica's severe climate postponed territorial claims until 1908, when Britain absorbed the northern part of the Antarctic Peninsula, known as Graham Land, as part of its Falkland Islands Dependencies. And a bit more, apparently: originally announced as encompassing the area "south of the 50th parallel of south latitude and lying between the 20th and 80th degrees of west longitude," the claim was modified in 1917 to exclude parts of the Argentinean and Chilean mainland north of the 58th parallel and west of 50 degrees (fig. 4.8).[22] The annexations that followed were similarly sluggish. New Zealand, France, and Argentina announced claims in 1923, 1924, and 1927, respectively; Australia carved out two large portions in 1933; and Norway and Chile helped themselves in 1939 and 1940.[23]

Ice cream torte, anyone? Antarctica's vaguely circular shape and the convergence of claim boundaries at the South Pole suggest a giant frozen desert consumed over the course of an evening by well-mannered guests with

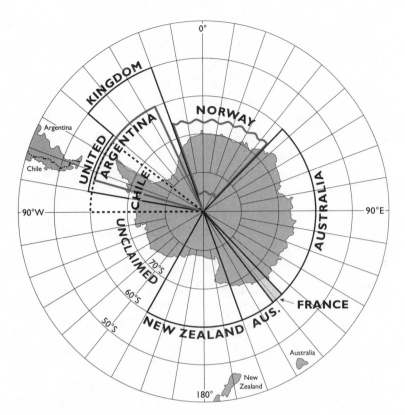

FIGURE 4.8. Seven countries claim slices of Antarctica.

diverse appetites and a shared reluctance to take the last slice. Although the
analogy is undermined by Argentinean and Chilean claims that overlap
each other as well as Britain's king-size portion, the map's straight-line
borders and rounded outer perimeters are typical of the sector principle,
whereby political cartographers anchor radial boundaries to coastal positions, islands, and landmarks on other continents.[24] In making the initial
cut, Britain used round-number longitudes to widen a slice tied to the South
Sandwich Islands on the east and the Antarctic Peninsula on the west. By
contrast, France played to the stereotype of a finicky eater by basing its
comparatively modest sliver on the explorations of Jules Sébastien César
Dumont d'Urville, who landed nearby on a rocky coastal island in January
1840. Bounded by meridians at 136° E and 142° E and named for Dumont
d'Urville's wife, Adélie Land separates Australia's two, much larger sectors,

claimed with help from Britain, which a decade earlier had asserted New Zealand's right to the Ross Dependency. The Brits found maps useful for visualizing possibilities as well as advertising claims.

Not so fast, cried Argentina and Chile, which viewed Britain's maps as an unseemly attempt to highjack their geographic birthright. In arguing that relative proximity was at least as important as precedence, the two South American claimants cited geologic links between the Andes and the mountains of the Antarctic Peninsula as well as an inherited right to Spanish territory west of the Tordesillas Line—reaching all the way to the South Pole, the papal meridian was the precursor of sector theory. Polar projections, on which straight-line meridians met at a point, helped claimants identify boundaries that were both logical and advantageous. In setting the western edge of its Antarctic territory at 90° W, Chile chose a round number that coincided with the western boundary of a mythical South American quadrant. By contrast, Argentina tied the western edge of its sector to the westernmost point along its mainland border with Chile but rounded the longitude outward to the next whole number.

In delimiting the eastern edge of its Antarctic claim, Argentina invoked what might be termed the expansive propinquity principle: if it's off our coast, even way off our coast (but not near someone else's coast), it's rightfully ours. Under this doctrine, claims and occupation by distant outsiders like Britain don't matter, even when other nations recognize British administration. Looking farther and farther eastward, Argentina has yearned to control the Falkland Islands (Islas Malvinas), the South Georgia Islands, the South Orkney Islands, and the South Sandwich Islands. The latter archipelago, where longitude ranges from 26°23' to 28°08' W, anchors Antártida Argentina to the meridian at 25° W, rounded outward to a multiple of five. Although Buenos Aires had announced a less expansive claim in 1927, its Antarctic sector was largely dormant until the mid 1940s, when the Peron government, in a fit of nationalism, promoted a puffed-up cartographic identity that defiantly persists in the country's schoolbooks and atlases—by law, all maps must include the Islas Malvinas and Antártida Argentina.[25] Most maps comply by including a small-scale inset at the lower right (fig. 4.9).

Sector theory was least appealing to tiny Norway, which opposed Russia's sector-based claims to a vast slice of the Arctic. Norway's interest in Antarctica reflected a substantial whaling fleet in the Southern Ocean as well as a passion for polar exploration. In early 1939 King Haakon VII

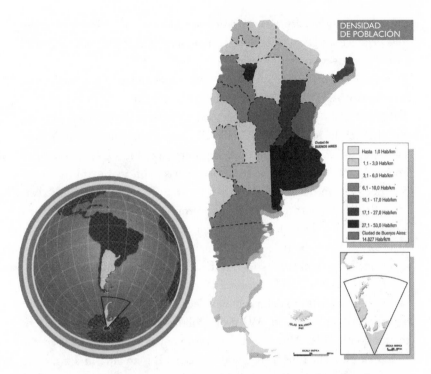

FIGURE 4.9. The logo of the Instituto Geográfico Militar, Argentina's national mapping organization, shows an Antarctic territory nearly as large as the Argentinean mainland (*left*). The logo appears several times in the agency's colorful national atlas, in which all thematic maps include a smaller-scale inset for the Antarctic sector, illustrated here by a map reporting a population density (*right*) of "less than one person per square kilometer" for the largely uninhabited territory.

proclaimed sovereignty "over that land which until now has lain unclaimed and which none but Norwegians have mapped and claimed."[26] Desperate to preempt claims by Germany, which was planning to explore the area by air, Norway named the area between Australian and British sectors Queen Maude Land and based its claim on a history of whaling in the region and several significant twentieth-century expeditions, including Roald Amundsen's heroic trek to the South Pole by dogsled in 1911. Willing to accept the Australian and British meridians as eastern and western borders, the king's advisors balked at delimiting specific northern and southern boundaries—hence the wavy line on figure 4.8. No point in endorsing a theory that gave Russia a distinct advantage in the Arctic.[27]

Overseas holdings without the expectation and expense of immediate occupation were especially attractive to Germany, which had lost its African colonies at the Paris Peace Conference. Undeterred by Norway's claim to Queen Maud Land, announced on January 14, 1939, Berlin proclaimed sovereignty over an area between 20° E and 20° W five days later, when German aircraft began taking aerial photographs and dropping medallions with swastikas.[28] Named Neuschwabenland after the ship that supported the seaplane sorties, the claim appeared on Nazi maps but died with the Third Reich in 1945. Hitler's Pacific ally, Japan, had whaling interests in the waters off Antarctica but never registered a claim.[29] During the 1930s South Africa had contemplated a sector based on its proximity to the Southern Ocean but situated in what became Queen Maude Land. Despite British support, the Pretoria government dropped the idea when Norway registered a wider and more convincing claim.[30]

How far from the pole can one anchor a sector? About 6,000 miles, according to geopoliticians in tiny Ecuador, which straddles the equator. Ecuador owns the Galápagos Islands, an archipelago with sufficient longitudinal breadth to delimit a four-degree slice of Antarctica (between 94°59'30" W and 98°59'30" W) if one follows these bounding meridians to the South Pole. Although the government never fully endorsed the idea, first published by army colonel Marcos Bustamante in a Brazilian newspaper in 1956, a similar assertion of Ecuadorian sovereignty by the country's Constituent Assembly 11 years later triggered an objection by Chile.[31]

Ecuador is not the only South American nation to covet a distant slice of Antarctica. Peru, Uruguay, and Brazil have mapped similarly theorized sectors, and a 1985 CIA map flattered the latter's geopoliticians by reporting a "Brazilian Zone of Interest" south of 60° S.[32] That Brazil had mapped an "informal claim" between 28° W and 53° W apparently made it noteworthy to the CIA, whose map also included the seven formal claims, none recognized by either the United States or Russia.

All claims have been on hold for a half century, thanks to the Antarctic Treaty System (ATS), negotiated in 1959 after the burst of scientific cooperation known as the International Geophysical Year, which ran from mid 1957 through the end of 1958. A short document as international agreements go, the ATS bans military activity and nuclear testing poleward of 60° S, opens the area to scientific exploration and other peaceful purposes, and provides for inspections as well as an annual meeting of the treaty's 28 "Consultative Parties," who have approved several important environmental safeguards.[33] While the agreement neither abolishes nor diminishes prior claims, it prohibits new or expanded declarations while the treaty is in force and also precludes using activities initiated under the ATS to either support or deny past or future territorial claims. While the seven claimants continue to advertise their paper territories, the closest thing to a map of effective occupation shows 37 year-round research stations maintained by 21 countries with no need to pay rent or ask permission.[34]

Like the lone slice of pie left at the end of a meal, the huge "unclaimed" sector caught the eye of several nonclaimants, notably Malaysia, situated due north (but on the far side of Indonesia), and India, still farther north and off to the west as well.[35] Needy nations surely, but not uniquely worthy. Rather than assert a poorly rationalized claim, easily contested by nations not so distant, their leaders embraced the principle of "common heritage," which would share Antarctica's untapped and largely unknown riches more widely, possibly under UN administration. An intriguing idea, no doubt, but if the ATS falls apart, don't expect the seven claimants to go quietly. Once inscribed on a map, territorial boundaries become an enduring argument for entitlement and redress.

CHAPTER
FIVE

....

DIVIDING
THE SEA

In much the same way that polar maps became an argument for slicing up Antarctica, nautical charts fostered seaward expansion of national sovereignty in the latter half of the twentieth century. Increased awareness of the oceans as a rich but limited resource encouraged a handful of maritime nations to proclaim exclusive rights to waters as much as two hundred nautical miles offshore. Other countries voiced alarm or pondered similar claims. Jurisdiction over near-shore territory had been a way of warning off smugglers and hostile warships, but the traditional territorial sea was a mere three nautical miles wide, the maximum reach of eighteenth-century artillery. Eager to avoid conflict and also seize an opportunity, coastal states settled on 200 nautical miles as an acceptable standard—as long as it focused on fish and oil. United Nations officials dubbed this new jurisdiction the Exclusive Economic Zone (EEZ) to distinguish its more limited rights from the prerogatives of the territorial sea, increased in the same frenzy to a uniform width of 12 nautical miles. EEZ boundaries began showing up on world maps to the surprise of Americans who never envisioned a maritime border with Venezuela, New Zealand, or Japan.

In exploring international efforts to divvy up a good part of the world's oceans, this chapter looks first at the definitions and delineations of the territorial sea, the EEZ, and other maritime territories. Like their dry-land counterparts, maritime claims require maps and legal codes, in this case a series of UN "conventions" as well as specific treaties between nations sharing a maritime border. As a few revealing examples demonstrate, fixing an offshore boundary can be troublesome, even when the only adjoiner is the High Sea—other nations, seeing their rights threatened by excessive claims, protest vigorously. In addition to examining maritime boundary controversies, this chapter looks at the value of islands and coastal irregularities in claiming jurisdiction over coastal waters, the possibility that sea level rise or erosion might extinguish the sizeable EEZs surrounding low-lying island nations, the global positioning system's role in telling would-be trespassers to keep their distance, and the emergence of the "Marine Protected Area" as a new frontier of environmental protection.

Marginal Imperialism

Maritime territory comes in different strengths and thicknesses, shown on maps as overlapping bands. Anchored to the low-water shoreline, each band extends seaward a fixed distance, with the wider zones conferring less control. The narrowest and most restrictive band is the territorial sea, a 12-nautical-mile zone within which a coastal country can enforce its statutes and regulate navigation. International law allows civilian and military vessels of all nations expeditious "innocent passage" through a territorial sea from one part of the High Seas to another—just passing through is fine, needless lingering isn't. Under UN rules, countries can elect a territorial sea narrower than 12 miles, but few do. In addition, nations can enforce their environmental, customs, and immigration laws within 24 nautical miles of the coastline. For most maritime countries this provision tacks a 12-mile "contiguous zone" onto their 12-mile territorial sea—the added buffer affords increased control over smuggling and waste dumping. Within 200 nautical miles of the coast, the EEZ adds the potentially lucrative right to control fishing, drill for oil, and mine the seabed.

Political cartographers delineate maritime zones by drawing overlapping circles centered along the low-water shoreline. The intersecting arcs of these circles define an envelope of points within 12, 24, or 200 miles of this "normal baseline." As figure 5.1 demonstrates, it's often sufficient to consider only circles centered on prominent headlands. To expedite delin-

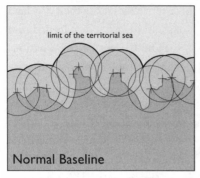
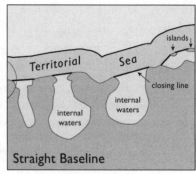

Normal Baseline

Straight Baseline

FIGURE 5.1. Key elements of normal (*left*) and straight (*right*) baselines.

eation, official mapmakers sometimes use "straight baselines" to simplify a highly irregular coastline, connect closely spaced coastal islands, and close off bays and wide rivers. Cartographic expediency reinforces national security insofar as the baseline separates a country's territorial sea from its "internal waters," where foreign ships have no right of innocent passage.

Because overly long straight baselines might unfairly enlarge a nation's internal waters, straight baselines raise the question, "How long is too long?" UN rules allow a closing line across the mouth of a bay as long as 24 nautical miles—twice the width of the territorial sea. Intersecting portions of a territorial sea anchored on opposing shorelines thus close off wider, more interior portions of estuaries like Delaware Bay. Although longer closing lines are allowed for "historic bays," the rules don't include a definition. A long-standing and continuous claim acknowledged by other nations, particularly those with big warships, seems sufficient; other circumstances invite complaint. The United States was quick to protest when Libya proclaimed a closing line 300 nautical miles long across the Gulf of Sidra in 1973.[1] To contest Libya's claim, the U.S. military held training exercises there. By contrast, the State Department has no qualms about China's claim to Bohai Bay, an appendage to the Yellow Sea enclosed by headlands 55 nautical miles apart.[2]

Today's 12-nautical-mile territorial sea reflects an increased desire for offshore sovereignty as well as a shared resistance to extravagant claims. A 1930 survey that identified 37 countries with a territorial sea found widths ranging from three to six miles (reported by 20 and 12 countries, respectively). In a 1951 survey that uncovered 67 claimants, 12 miles was the greatest width (claimed by 3 countries) and three miles was the most common (endorsed by 41 nations).[3] In a 1958 survey of 57 claimants, the three-mile sea was still the most frequent (favored by 23 nations), but the number of 12-mile seas had tripled (from 3 to 9). Particularly ominous were territorial seas 30 and 200 nautical miles wide, claimed by Chile and El Salvador, respectively. Small countries eager for greater control of offshore activities favored a large territorial sea while large nations with powerful navies preferred a narrow standard less likely to hamper military maneuvers in remote regions. In the 1960s, most nations expanded their territorial sea from three to twelve miles. While the United States stubbornly held onto its three-mile limit, many Latin American countries announced a 200-mile jurisdiction.[4]

Twelve miles did not become the world standard until 1982. Representatives from more than 160 countries had worked since 1973 to replace four earlier treaties with the United Nations Convention on the Law of the Sea (UNCLOS), not officially effective, or "entered into force," until 1994. One of these obsolete agreements was the 1958 Convention on the Territorial Sea and Contiguous Zone, which reaffirmed a right of innocent passage and endorsed a contiguous zone 12 miles wide. Despite its title, the treaty sidestepped a uniform territorial sea largely because of American and British resistance. The United States had ratified the 1958 treaty, but conservative senators stubbornly resisted the 1982 agreement.[5] Equally obstinate are Peru and a few other countries, which continue to claim a 200-mile territorial sea.[6]

Convinced that its traditional territorial sea was fully adequate, Washington ignored the permissible maximum for six years. The decisive factor was an increased presence of Soviet spy ships just outside the three-mile limit. When Ronald Reagan signed Presidential Proclamation 5928 in late 1988, the United States became the 105th country to declare a 12-mile territorial sea.[7] (A nation can implement a provision of a treaty without becoming a signatory.) Eleven years later another long-overdue presidential proclamation added the 12-mile contiguous zone allowed under the 1958 treaty. The action doubled the jurisdiction within which the Coast Guard can board and search foreign vessels suspected of violating U.S. environmental and fishing regulations.[8]

American resistance to UN solutions was most apparent in Washington's go-it-alone approach to the 200-mile EEZ. In 1976, while Law of the Sea negotiators were hashing out a compromise between developing nations eager for a 200-mile territorial sea and wealthier countries content with a more limited jurisdiction, Congress established a 200-mile Fisheries Conservation Zone (FCZ) as part of the Magnuson Fishery Conservation and Management Act. Intended to protect a threatened resource as well as reinvigorate a declining industry, the law set up regional fisheries management councils to determine how many of what species might be caught within the zone and who could catch them—when up and running, the councils cut the foreign share of the FCZ catch from 61 percent in 1981 to just 1 percent in 1991.[9] In March 1983, three months after the new UN agreement was ready for signatures, President Reagan replaced the FCZ with an EEZ that endorsed many of the UNCLOS provisions.[10] The United States would have its EEZ but not under UN rules.

What soured ratification was the International Seabed Authority, a new UN bureaucracy set up to regulate deep seabed mining.[11] According to James L. Malone, Reagan's assistant secretary of state for oceans and international environment and scientific affairs, the Authority "would have . . . the unprecedented power to redistribute wealth on an international scale." In a further affront to American sovereignty—and to conservative Senate Republicans in particular—the treaty would "jeopardize private access to seabed resources" and "create other undesirable precedents, such as the mandatory transfer of technology and distribution of funds to national liberation movements."[12] A 1994 amendment to UN regulations on seabed mining failed to overcome Senate resistance, and the UNCLOS agreement remained unratified for more than a quarter century.

Thirty-four countries, give or take, enjoy additional offshore privileges. Where the adjacent continental shelf extends beyond its EEZ, a coastal nation is entitled to jurisdiction over the "seabed and subsoil" on the continental shelf as far as 350 nautical miles beyond its baseline.[13] Delineating the shelf's outer limit is more complex than anchoring an envelope of circles to a baseline: coastal nations may not merely assume the continental shelf extends an extra 150 nautical miles beyond their EEZ. Because technical criteria require a significant investment in seafloor mapping, UNCLOS Article 76 encourages claimants to consult with the UN Commission on the Limits of the Continental Shelf.[14] With no formal powers to confirm or deny a claim, the Commission has the ambiguous role of facilitator and "legitimator"—UN technocrats can study a claimant's "submission" and issue a "recommendation," but they have no enforcement powers.[15] As with other kinds of maritime boundaries, the claimant announces its jurisdiction unilaterally, and countries that consider the claim excessive are free to

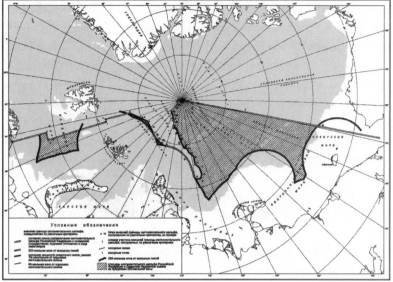

FIGURE 5.2. Russian Federation's claim to the continental shelf in the Arctic Ocean beyond the 200-nautical-mile EEZ. From the Commission on the Limits of the Continental Shelf Web page for the Russian Federation's submission, http://www.un.org/Depts/los/clcs_new/submissions_files/rus01/RUS_CLCS_01_2001_LOS_2.jpg.

protest. Although not definitive, the Commission's guidance reduces the likelihood of excessive claims and groundless protests.

The Commission received its first submission in 2001, from the Russian Federation.[16] Separate maps delineated claims in the Arctic and Pacific Oceans. The Arctic map (fig. 5.2) shows two zones: a large area in the Barents Sea due north of Murmansk and a still larger zone farther east in the Central Arctic Ocean. The latter territory extends to the North Pole, its straight-line eastern boundary confirming Russian reverence for sector principles as well as a questionable argument that the continental shelf reaches beyond the North Pole. Its irregular western boundary reflects undersea terrain. According to Russian geoscientists, the Alpha-Mendeleev Ridge, a submerged mountain range rising tens of thousands of feet upward from the deep seabed, is a natural prolongation of their continental shelf, and thus beyond the jurisdiction of the International Seabed Authority.

Canada and the United States disagreed. The U.S. State Department conceded that the ridge system was "the surface expression of a single continuous geologic feature," but argued that, as a comparatively recent volcanic formation, "it is not part of any State's continental shelf"—as the Law of the Sea defines the shelf, thick sedimentary rocks linked to the continent count, but Johnny-come-lately volcanic outliers don't.[17] UN officials studied the Russians' maps, consulted numerous geoscientists, concluded the case was weak, and requested a revised submission. (In accord with the Commission's rules, the full text of its response is secret.) Although Russia has yet to submit new data, its dramatic planting of a titanium flag on the ocean floor beneath the North Pole in August 2007 indicates a continued interest in controlling the Arctic seabed, potentially rich in oil.[18]

Russia and the United States might disagree over mining rights at the North Pole, but in 1990 the two nations had quietly settled their line across the polar sea—at least its trajectory, if not its extent—by precisely delimiting the meridian established in 1867, when Secretary of State William Seward purchased Alaska for $7.2 million. As described in the new agreement, the northern portion of the boundary "extends north along the 168° 58' 37" W meridian through the Bering Strait and the Chukchi Sea into the Arctic Ocean as far as permitted into the Arctic Ocean under international law."[19] Seward and Russian ambassador Eduard Stoeckl had agreed on a meridian equidistant between Big Diomede Island and Little Diomede Island, but the 1990 treaty gave the line a specific longitude. As shown in figure 5.3, the United States also gained the "Eastern Special Area" just east of the agreed meridian. Without the treaty, this area would be part of the Russian EEZ. To avoid wasting maritime territory on the wrong side of the 1,600-nautical-mile boundary, which runs southwestward across the Bering Sea, the United States received three Eastern Special Areas while the USSR gained a Western Special Area.

America's other Arctic maritime boundary, in the Beaufort Sea, remains unresolved. The United States favors an equidistance line, which assigns offshore territory to whichever nation has the closest baseline, while Canada prefers the sector principle, which would extend the land boundary between Alaska and the Yukon poleward along the same meridian.[20] As shown in figure 5.3, a northward bulge in the Alaskan shoreline pushes the equidistance line eastward, to the benefit of the United States. Canadians argue that the United States endorsed the sector principle in its 1990 treaty with the former Soviet Union—why not give us the same consideration, they ask? Americans reply that the U.S.-Soviet agreement was a negotiated settlement with benefits for both parties, not an affirmation of sector

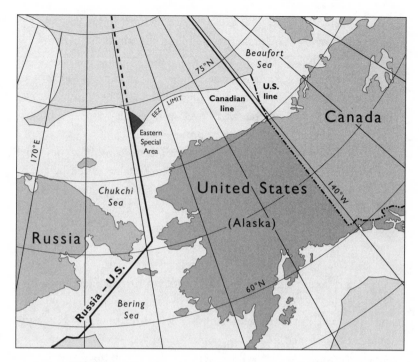

FIGURE 5.3. U.S. maritime boundaries in the Arctic Ocean.

theory. The 1867 meridian was intended merely to divide islands in the Arctic Ocean among the two countries and had nothing to do with maritime territory. The dispute is decades old, but oil reserves in the Beaufort Sea are an incentive for resolving the impasse, perhaps with a compromise boundary.[21] A trade-off of fish for oil might help the two countries resolve their other maritime boundary disputes, in both the Atlantic and Pacific Oceans.[22]

Distant Neighbors

EEZs underscore the importance of islands, especially those far offshore. To qualify for an exclusive economic zone, an island nation or insular possession need only be a "naturally formed area of land" at least partly above water at high tide and able to "sustain human habitation or economic life of [its] own"—size of the dry, habitable area is not an issue.[23] In principle, a tiny island no larger than a few city blocks could control a territorial sea and an EEZ of 450 and 125,000 square nautical miles, respectively. France, which owns numerous islands around the world, has the world's second largest EEZ, not far behind the United States.[24] Isolation from other countries' islands is particularly valuable because a good chunk of an island's otherwise exclusive 200-mile buffer might lie closer to another nation's baseline. Where potential EEZs overlap, maritime territory is usually divided according to the equidistance principle.

Because the Law of the Sea gives countries leeway over how they construct their baselines, maritime boundaries are usually settled by treaty. In addition to resolving potential disputes over closing lines, tidal data, low-water shorelines, and the relative merits of different charts and map projections, treaties simplify maritime boundaries by reducing them to a straightforward list of points, each identified by its latitude and longitude referenced to a specific, mutually acceptable geodetic datum.[25] Aside from noting that different datums often yield different latitudes and longitudes for the same point, I won't delve into the mechanics of coordinate systems, map projections, and other heady stuff best left to textbooks on geodesy.[26] A boundary published as a list of points is easily and reliably plotted on paper or electronic charts. Each segment is a shortest-path, "geodesic" line between successive points. Coordinates work well because a skilled navigator with a GPS unit and an accurate chart should have no trouble determining his or her position relative to the boundary.

A typical example is the maritime boundary between the United States and Venezuela, negotiated in the late 1970s.[27] Along this border neither country's EEZ is wider than 177 nautical miles. As described in figure 5.4, the boundary is a string of 22 "turning points" referenced to a mutually acceptable datum. Starting at point 1, a "tripoint" equidistant from the Netherlands Antilles, the United States, and Venezuela, it runs southwestward along a course equidistant between St. Croix (the largest of the U.S. Virgin Islands) and Isla Aves (a distant outlier of Venezuela). At point 8, where proximity to Puerto Rico becomes relevant, the boundary takes a more southerly route to point 11, where islands north of the Venezuelan mainland become influential. It then proceeds westward to point 22, where an arrow shows its continuation "along an azimuth of 274.23 degrees" (according to the agreement) to a yet-to-be-determined end point involving an unnamed "third State"—the map suggests either the Dominican Republic or the Netherlands. Countries can negotiate an equidistance line

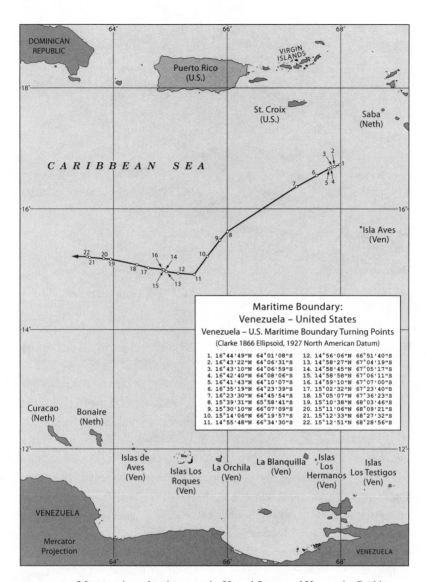

FIGURE 5.4. Maritime boundary between the United States and Venezuela. Grid lines shown are roughly 120 nautical miles apart.

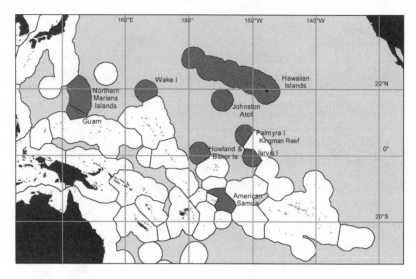

FIGURE 5.5. EEZ waters of the western Pacific, with U.S. maritime territories named and highlighted.

between their two territories but cannot bilaterally commit a third nation to a specific tripoint.

The importance of islands in claiming maritime territory is most apparent in the Western Pacific, where EEZs swallowed a huge chunk of the High Seas (fig. 5.5). New neighbors eager for agreed boundaries sometimes used the negotiations to resolve other disputes. In 1980, for instance, talks between the Cook Islands and the United States over a maritime boundary for American Samoa inspired Washington to renounce claims to Manihiki, Penrhyn, Pukapuka, and Rakahanga.[28] These four tiny islands are part of an archipelagic nation of 22,000 residents distributed over 16 islands in the South Pacific midway between Hawaii and New Zealand—with a land area only 1.3 times that of the District of Columbia, Cook Islanders enjoy a maritime territory nearly three times the size of Texas. The U.S. claim was clearly obsolete: American whalers had visited the four atolls in the nineteenth century, and the 1856 Guano Islands Act had authorized U.S. citizens to take possession of ungoverned islands with substantial deposits of guano (dried bird droppings), a prized fertilizer. Like many questionable territorial claims, this one was not worth arguing over.

Another 1980 treaty, concluded with New Zealand, fixed the maritime boundary between American Samoa and its western neighbor Tokelau and confirmed U.S. sovereignty over Swains Island—at least for now.[29] Tokelau is a three-island protectorate that New Zealand inherited from Great Britain. Swains Island is really a colony of American Samoa, itself a colony of the United States; also known as Olohega, it became a U.S. possession because Eli Hutchinson Jennings, Sr., a Yankee entrepreneur, established a coconut plantation there after purchasing the island in 1856 from a British explorer who claimed ownership. The two dozen Swain Islanders speak Tokelauan, and the 1,500 Tokelauans under New Zealand administration are reluctant to sever ties with their cultural kinfolk. In 2006, as a prelude to becoming a self-governing state in "free association" with New Zealand, Tokelauan leaders drafted a constitution that claims "at the dawn of time the historic islands of Atafu, Nukunonu, Fakaofo, and Olohega were created as our home."[30] Although Totelauan voters failed twice to give the draft the required two-thirds majority, the CIA's online World Factbook, which lists 10 unresolved territorial disputes with the United States, warns that "Tokelau included American Samoa's Swains Island among the islands listed in its 2006 draft constitution."[31]

Tokelauans have more to fret about than an American flag over Swains Island. With their highest point a mere five meters (16 feet) above water line, they're exceptionally vulnerable to storm surge, tsunamis, and sea level rise.[32] The Intergovernmental Panel on Climate Change expects the world's oceans to rise between eight inches and two feet over the current century—assuming no accelerated melting of glaciers in Antarctica and Greenland.[33] Too poor to build a substantial sea wall, Tokelau is considered more vulnerable to climate change than the Republic of the Maldives, where the highest elevation is only 2.5 meters (eight feet).[34] A group of more than a thousand coral islands, about 200 of which are inhabited, the Maldives is a popular tourist destination, hailed as "the world's flattest nation."[35] To survive, the country is supplementing the six-foot sea wall around Male, its capital, by building a four-square-mile artificial island a short boat ride away. Named Hulhumalé, this huge reclamation project will provide a refuge—temporarily, at least—for most of the Maldives' 370,000 residents.

Whether Hulhumalé can save the Maldives' disproportionately large EEZ is debatable. And so is the question of whether Tokelauans evacuated to New Zealand will retain rights to their own EEZ after their last island disappears beneath a rising high-tide line. While the UN Convention on the Law of the Sea lets coastal nations build artificial islands within their maritime territory, Article 60 clearly states that these structures "have no territorial sea of their own, and their presence does not affect the delimitation of the territorial sea, the exclusive economic zone or the continental shelf."[36] What's more, because an island must, by definition, be "above water at high tide," and because "rocks which cannot sustain human habitation or economic life of their own shall have no exclusive economic zone or continental shelf," permanent inundation should extinguish any claim to an EEZ, not to mention the survivors' status as a coastal state.[37] Even so, the Law of the Sea could be revised to accommodate events largely unforeseen decades earlier, or the International Court of Justice, in The Hague, could call for compensation from countries that contributed to climate change.[38] It's easy to see why sea level rise remains a touchy subject for countries hooked on oil.

Restrictions and Enforcement

Delimiting an EEZ is merely an early step to controlling what happens within its boundary. In managing its fishery, for instance, a nation can limit the size of the catch, set quotas, prohibit certain types of nets and lines, protect threatened species, exclude vessels lacking an entry permit, levy taxes or catch-based fees, impose fines, and even seize an offending vessel and its catch.[39] In addition to boarding and inspecting nonmilitary vessels within its EEZ, a country can use onboard observers and electronic catch-monitoring systems to monitor compliance. And when migratory species ignore EEZ boundaries—nobody bothered to tell the fish—conservation treaties and EEZ-related enforcement offer supplementary protection.

Within its EEZ a nation can also control dumping as well as regulate the transport of nuclear waste and other exceptionally hazardous materials. Although these powers received little attention in 1982, the mandate to protect and preserve the marine environment has broad implications. In 2004, for instance, Chile refused to let the U.S. Department of Energy ship a decommissioned nuclear reactor around Cape Horn from California to South Carolina for burial.[40] Environmental protection also seems a convenient justification for banning single-hull oil tankers, involved in several disastrous spills. As the Law of the Sea is evolving, the EEZ is not just about fish and seabed mining. Especially contentious for military powers like the United States is an emerging consensus in Asia and Latin

America that military operations, surveillance, and hydrographic surveying are inappropriate in the EEZ without the coastal country's approval.[41]

Effective (and fair) enforcement requires conspicuous publication of maritime boundaries. Advertising the coordinates of turning points is a promising approach because electronic navigation systems can help captains anticipate and avoid unintentional intrusions. Look out, though, for unresolved or disputed boundaries—putting a line on a chart is no guarantee a maritime neighbor will accept your interpretation, as the British forces occupying Iraq discovered in March 2007, when the Iran's Revolutionary Guard detained 15 British sailors for 12 days for violating Iranian waters.[42] Britain's Defence Ministry produced maps showing that the sailors had been captured within Iraqi waters, but because Iraq and Iran never had an agreed maritime boundary, this evidence was not conclusive.[43]

When negotiations fail, arbitration is useful, especially when neighbors are eager to get on with fishing or mining. In the early 1980s, for instance, Canada and the United States resolved their boundary in the Gulf of Maine by deferring to the International Court. The dispute included Georges Bank, a rich fishing grounds off Cape Cod over which the United States wanted exclusive jurisdiction. Not surprisingly, the justices elected a compromise boundary: Canada got a piece of Georges Banks, and neither nation was greatly pleased or disappointed.[44]

Expanded jurisdiction over territorial and EEZ waters, coupled with the need to tailor regulations to specific species and threats, requires charting of smaller, more focused zones called Marine Managed Areas (MMAs) or Marine Protected Areas (MPAs).[45] Less important than the bureaucratic distinction between an MMA and an MPA is the exact delineation of zones subject to specific regulations.[46] For example, the National Marine Fisheries Service, a division of the National Oceanic and Atmospheric Administration (NOAA), supports its Harbor Porpoise Take Reduction Plan with MPAs called "closure areas." The Web map in figure 5.6 describes the Northeast Closure Area, where gill nets are banned from August 15 through September 13. Designed so that fish can swim in but can't back out, gill nets can inadvertently trap harbor porpoises, a protected species, like all marine mammals. Harbor porpoises migrate along the east coast, and in late August and early September they concentrate in this area.[47] When I looked, NOAA's harbor porpoise Web site posted restrictions for different times or other fishing gear for additional closure areas,

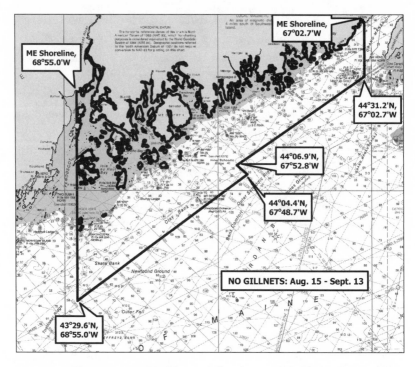

Inside the map:

ME Shoreline, 67°02.7'W

ME Shoreline, 68°55.0'W

44°31.2'N, 67°02.7'W

44°06.9'N, 67°52.8'W

44°04.4'N, 67°48.7'W

NO GILLNETS: Aug. 15 - Sept. 13

43°29.6'N, 68°55.0'W

FIGURE 5.6. Upper portion of the "Outreach Supplement to the Harbor Porpoise Take Reduction Plan (HPTRP): Northeast Closure Area." Annotation warns that the greatly reduced map is not intended for navigation. From NOAA Northeast Regional Office Web site, http://www.nero.noaa.gov/prot_res/porptrp/HPTRPNEClosure.pdf.

some wholly or partly outside the territorial sea.[48] As figure 5.6 illustrates, point coordinates provide a concise description of MPA boundaries, readily added to NOAA nautical charts.

Originally intended to safeguard sailors, not fish, nautical charts also protect underwater cables and pipelines, which might be damaged by anchors or draglines. Dashed boundaries printed in a distinctive magenta also delineate restricted areas around military bases and areas reserved for military testing or target practice. For example, a chart for coastal waters near Kittery, Maine (fig. 5.7), reveals a cable corridor (*top left*) and two zones cross-referenced to the *Code of Federal Regulations*, which lists boundary coordinates and specific restrictions. In Restricted Area 334.50, for instance, "All persons, vessels and other craft, except those vessels under the supervision of or contract to local military or naval authority, are prohibited from entering the restricted areas without permission from the

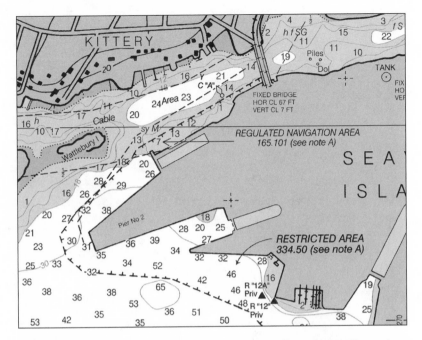

FIGURE 5.7. Cable corridor and restricted areas near the Portsmouth Naval Shipyard, Kittery, Maine.

Commander, Portsmouth Naval Shipyard or his/her authorized representative."[49] By contrast, in Regulated Navigation Area 165.01, the Coast Guard enforces a five-miles-per-hour limit speed.[50] As the nautical chart became a prohibitive map as well as a navigation tool, restrictive boundaries joined the more traditional pictorial symbols used to warn sailors away from wrecks, submerged rocks, and dangerous shoals.

CHAPTER

SIX

....

DIVIDE AND
GOVERN

Jurisdictional boundaries affect how we live our lives, respect our neighbors, and pay for public services. Adjoining states can differ markedly in tax rates and the types of taxes levied as well as in government services ranging from health care and higher education to wildlife conservation and environmental protection. Some states have both sales and income taxes, some have one but not the other, and some have neither. Real estate tax codes and the array of regulations affecting businesses large and small vary from state to state as do judicial codes and the powers and responsibilities delegated to local governments. Some counties and towns ban the sale of alcoholic beverages altogether, while others let supermarkets sell beer and wine, and a few let bars stay open until 4 a.m. Neighboring municipalities often have different approaches to enforcing traffic laws, repairing roads, and inspecting rental housing, and civic pride and community attitudes toward efficiency and openness further heighten contrasts among local jurisdictions. In addition, local political boundaries provide a ready framework for overlays of conservation districts, regional transportation authorities, and similar administrative units.

This chapter explores the variety of jurisdictional boundaries found in the United States, with a focus on their cartographic treatment and diverse impacts. It reflects a personal fascination with local boundaries that began in the early 1970s, when I had what Marge and I call my "ice accident." We were living in a suburban town outside Albany, New York, a bastion of machine politics. Albany's CEO was Erastus Corning II, who became mayor in 1941 at age 32 and was reelected 10 consecutive times. The patrician great-grandson of Erastus Corning I, who amassed a fortune in various businesses and organized what became the New York Central Railroad, Erastus II owned an insurance company that was the sole bidder on county insurance contracts during most of his tenure as mayor—not an overt conflict of interest because this was county (not city) business. Corning kept ward heelers happy by employing their elderly friends and family members, particularly in the highway department. That city streets often went unplowed became expensively apparent one day when we drove into the city along a road our suburb had cleared promptly after a snowfall the day before. As I crossed the city line, the road veered to the left, the surface changed abruptly from asphalt to ice, and my car slid into a wooden fence. While the sympathetic homeowner helped me pry the fender away from the tire, a city truck drove by, with three senior citizens shoveling salt out the back. From that day forward I've been particularly wary of municipal boundaries during winter.

Slice and Dice

Subnational jurisdictions address many functions, some quite narrow. School districts, for example, deliver a specific service—public education—and in many areas their boundaries are often vague and surprisingly changeable. Like fire districts, legislative districts, and "special districts" configured to provide drinking water, sewers, or street-lighting, they have their own maps, available for inspection if you ask the right person. By contrast, states, counties, and towns have broader responsibilities and comparatively stable boundaries, eminently appropriate for the large-scale topographic maps provided by the U.S. Geological Survey as well as many road maps. Federal cartographers recognize the varying importance of these lines of separation with a hierarchy of boundary symbols. Figure 6.1, taken from the back of a USGS topographic map printed in 1909, contrasts the thicker, more prominent lines used for state borders with the progressively thinner or more intricately dashed boundary symbols

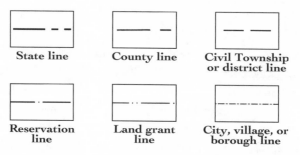

State line   County line   Civil Township
                             or district line

Reservation   Land grant   City, village, or
line   line   borough line

FIGURE 6.1. Principal boundary symbols used by the U.S. Geological Survey in the early twentieth century.

for counties and minor civil divisions. By convention, all USGS boundary lines are dashed, to distinguish them from symbols for roads, rivers, and shorelines. In general, states are subdivided into counties, which are split into civil townships or districts. Cities typically cover only part of a county, and villages or boroughs are smaller, more highly focused centers of population. In New York State, where I live, an incorporated village is also part of a town, which is also part of a county. The array of federal boundary symbols includes land grants, generally shown only in the West, and reservations such as military bases, national parks, and tribal lands. Section lines (see chapter 2) and other land-line boundaries are sometimes shown, often in red.[1]

Although property lines appeared on maps long before political subdivisions, administrative cartography was closely tied to hierarchical strategies for building roads, collecting taxes, and defending territory. Chinese military maps from the second century BC as well as pre-Christian Roman maps delineate local jurisdictions, and Christopher Saxton included subdivision boundaries on county maps in his 1579 *Atlas of England and Wales*, hailed by map historians as the first national atlas.[2] In the late twentieth century the shift from paper to electronic maps coincided with the rapid growth of government at all levels, and boundary data became a basic element of the digital cartographic databases that comprise the modern national map. In 1995, when the Federal Geographic Data Committee rolled out plans for a massive coordination effort called the National Spatial Data Infrastructure, administrative data was one of seven "framework" themes.[3] Because up-to-date boundaries are essential for accurate census tabulations, the U.S. Census Bureau was named the responsible agency.

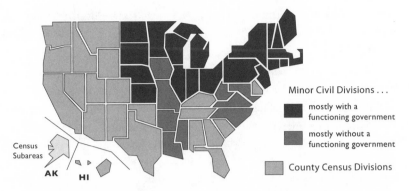

Minor Civil Divisions ...

■ mostly with a functioning government

■ mostly without a functioning government

▢ County Census Divisions

FIGURE 6.2. Types of county subdivisions, by state. Compiled from U.S. Census Bureau, *Geographic Areas Reference Manual*, 2005.

Bureau officials coined the term "census geography" to describe the intricate hierarchy of boundaries used in tabulating census returns. The fundamental geographic unit recognized in the Constitution is the state, which legislatures have sliced into parishes in Louisiana and counties in 48 other states. Although a few Cajuns insist otherwise, Louisiana's parishes are really counties. By contrast, sparsely populated Alaska is split among 16 "boroughs," which are functional governments, and 11 "geographical census areas," which aren't—in the Alaskan constitution these 11 areas comprise a single "Unorganized Borough." Elsewhere, a state's counties are divided into either "minor civil divisions" (MCDs), which are legally defined and generally stable, or "county census divisions" (CCDs), which are oriented toward a particular community and have visible, generally stable boundaries.[4] A state can have one or the other, not some of each. Alaska is the exception; lacking counties, it has "census subareas."

Census geography has its own geography. As shown in figure 6.2, MCDs are dominant in the Northeast and Midwest, while CCDs are the rule in the West and Southeast.[5] In nine states, mostly south of the Mason-Dixon Line or west of the Mississippi River, MCDs typically have no government functions. Maryland, for instance, divides its counties into "election districts" or "assessment districts," neither of which are local governments. In the other 19 MCD states, the vast majority of county subdivisions are actively functioning local governments. For example, all but one of New Hampshire's 222 towns operates as a local government. The exception is Livermore, in largely rural Grafton County; dissolved by the state legislature in 1951, Livermore has the right to reactivate—unlikely, though, because the town apparently has no permanent residents.[6] New Hampshire's other 24 nonfunctioning county subdivisions are classified as grants (8), locations (4), purchases (6), and townships (6). The state also has 13 incorporated cities, which are separate MCDs, independent of any town. Not all oddly named MCD types are dysfunctional: Maine's 36 plantations have active governments—unlike its single "gore" and 35 "unorganized territories," which have few, if any, permanent residents.

What could be called the Town Government Belt stretches from New England across the Midwest to the foothills of the Rockies. This pattern of local, subcounty control is no accident. Settlers who migrated westward from New England and the Middle Atlantic states in the late eighteenth and early nineteenth centuries roughly followed parallels of latitude, and so did their compatriots farther south. The result was a de facto extension of the Mason-Dixon Line, which heightened the contrast between "free" and "slave" states and helped precipitate the Civil War. Contemporary landscapes show traces of a "New England Extended" in their place names, house types, and street patterns,[7] and figure 6.2 confirms that transplanted Northeasterners brought along their preferred approach to local government.

A profound political imprint is also apparent in larger-than-average proportions of women elected to local office in states with actively functional MCDs. The connection seems clear: an expanded role for local government enlarges the number of elected positions, which in turn increases opportunities for female candidates. A map of the female percentage of elected officials in local government for 1992, the last year our quinquennial Census of Governments collected relevant data, shows only seven states with a distaff share above 30 percent (fig. 6.3).[8] All seven have active town or township governments—Michigan, Pennsylvania, and a few other states call their MCDs "townships." New Hampshire, which registered the highest rate (39.9 percent) and is famous for its open town meetings and inclusive boards of "selectmen"—nomenclature is slow to catch up—had more elected local posts than any single state in the Southeast. Although female office-holders have become more prominent since the early 1990s, their underrepresentation in the South and the Great Plains underscores the impact elsewhere of politically functional county subdivisions.

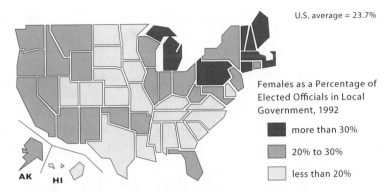

U.S. average = 23.7%

Females as a Percentage of
Elected Officials in Local
Government, 1992

■ more than 30%

▨ 20% to 30%

☐ less than 20%

AK
HI

FIGURE 6.3. Females as a percentage of elected officials in local government, 1992.
Percentages are based on office holders identified as either male or female in the 1992
U.S. Census of Governments. Sex was not identified for 14 percent of local officials.

Fluid Boundaries

Recent population growth in the Sunbelt has heightened the contrast between the Town Government Belt and the rest of the country. In regions
where county subdivisions don't provide government services, state law often gives new communities the options of incorporation and annexation.[9]
If there's no city nearby, incorporation might be the only option. But even
where a neighboring municipality could provide sewer connections or enhanced law enforcement, some communities choose incorporation over
annexation because a city or incorporated village can resist annexation
more readily than an unincorporated locality. Protective incorporation, as
it's called, might reflect fear of a big city's bigness, white aversion to having
a black mayor, or the desire of well-to-do homeowners to keep their taxes
low or control how the money is spent. Independence is a strong motive
for local leaders eager to become local office-holders, and once established,
incorporated places tend to remain independent. Despite obvious econo-
mies of scale, amalgamation of two or more existing local governments
has been rare in the Northeast and Midwest since the 1920s because local officials are reluctant to vote themselves out of office. The distinctive
identity that geographers call "sense of place" can be important too, as in
the small cities of Hamtramck and Highland Park, Michigan, collectively
surrounded by Detroit.[10] Despite devastating job losses in the automobile industry and severe fiscal challenges, neither the Polish stronghold of

Hamtramck nor its largely African American neighbor Highland Park is eager to amalgamate with Detroit—or with each other.

Race plays a critical role in municipal annexation in the American South, where small white cities have been deliberately configured, or "underbounded," to exclude black neighborhoods.[11] Although white cities routinely get away with annexing only white suburbs, disconnecting nonwhite neighborhoods is not an option, as Alabama discovered in 1961. Four years earlier the state legislature declared that "the boundaries of the City of Tuskegee in Macon County are hereby altered, rearranged and redefined so as to include within the corporate limits of said municipality all of the territory lying within the following described boundaries, and to exclude all territory lying outside such boundaries."[12] An attached metes-and-bounds description reshaped the perfectly square boundary of a city of about 7,000 residents, nearly 80 percent African American, into a 28-sided polygon (fig. 6.4) with only four or five registered black voters—clearly a ploy to preserve a white majority. Notably missing was the campus of the Tuskegee Institute, a private black college. While only a third of the 600 registered voters within the old boundary were black, the Civil Rights Movement was gaining momentum, and a black takeover seemed inevitable if not imminent. Although states have the right to redefine their municipalities, the U.S. Supreme Court ruled unanimously that Alabama had violated the Fifteenth Amendment, which bans racial discrimination in voting. Justice Felix Frankfurter, who wrote the decision, condemned the contorted boundary as "uncouth."[13] By late 1964, Tuskegee's new biracial government had started paving streets in black neighborhoods.[14]

States differ in their constraints on municipal annexation, including the requirement that added territory be contiguous. All states with unincorporated land that might be up for grabs have annexation statutes that demand contiguity of some sort, but wording varies.[15] Oklahoma, for instance, requires that the acquisition merely be "adjacent and contiguous"[16] to the expanding municipality, while a wordier Florida law demands "that a substantial part of a boundary of the territory sought to be annexed by a municipality is coterminous with a part of the boundary of the municipality."[17] Interpretation is left to the annexers, the annexees, and the courts. A few cities have stretched the notion of contiguity, both literally and figuratively, with "long lasso" or "strip" annexation, whereby a city expands its tax base by roping in land several miles away with a narrow strip of land included solely to provide contiguity. There's a limit, though, as

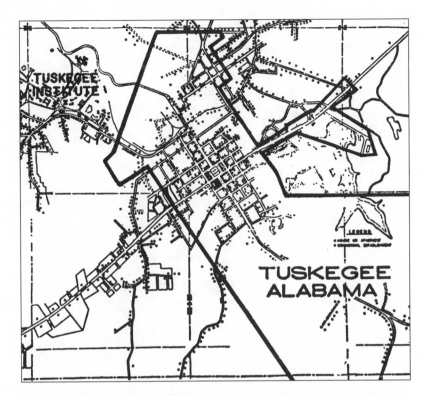

FIGURE 6.4. Map accompanying the 1961 Supreme Court ruling on Tuskegee, Alabama. According to the caption, the square shows the previous city boundary, and "the irregular black-bordered figure within the square represents the post-enactment city."

Seminole, Oklahoma, discovered in 2004, when the state's highest court balked at the city's attempt to annex land 10 miles away along a strip only three feet wide.[18] Seminole's ploy was perhaps understandable after earlier court decisions had condoned corridors 177.5 and 67 feet wide. Although the courts have become reluctant to say how narrow is too narrow, the Colorado legislature addressed the issue by banning connections longer than three miles.

Despite judicial reservations, weird shapes persist. Particularly ironic is Tuskegee, Alabama, whose current boundary is far more irregular than the municipal gerrymander struck down by the U.S. Supreme Court in 1961 (fig. 6.5). Eager to add taxable land, the city annexed a "dog track property" 14 miles out of town via U.S. Highway 80, a county road, and a segment of Interstate 85.[19] To satisfy the state's contiguity requirement, the city annexed the right-of-way. Wary of city control of the highway, an

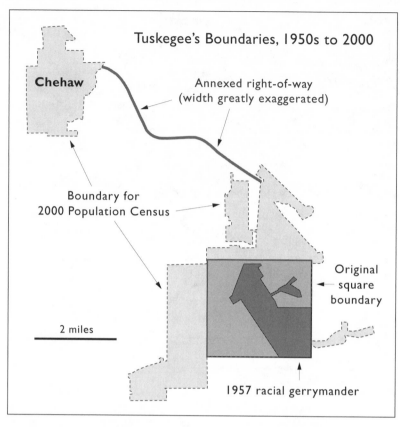

Tuskegee's Boundaries, 1950s to 2000

Chehaw

Annexed right-of-way
(width greatly exaggerated)

Boundary for
2000 Population Census

Original
square
boundary

2 miles

1957 racial gerrymander

FIGURE 6.5. Tuskegee, Alabama, in 2000, compared to the city limits before and after the 1957 racial gerrymander.

adjoining landowner sued, and the case eventually went to the Alabama Supreme Court, where five out of nine justices agreed that the connective corridor provided the needed contiguity—because the annexation statute was imprecise, the court accepted the dictionary definition of contiguity as "touching," even at a single point.[20] But two years later, a slightly different mix of justices—again by a slim majority—denounced the "spider-web effect" of two annexations proposed near Birmingham.[21] Although Alabama's high court simultaneously "overruled" its earlier interpretation, the outlying zone, known locally as Chehaw, remains attached to Tuskegee by an umbilical boundary readily apparent on Census Bureau maps.

Contiguity is hardly a prerequisite for efficient city boundaries. All state annexation statutes require it, partly to ensure that cities can properly service their new territory and partly to prevent greedy municipalities from cherry-picking the county tax base. Even so, meticulously contiguous annexation can leave islands of unincorporated land inside the city boundary. Laws in 12 states explicitly ban these enclaves, which can be difficult for county governments to service.[22] And because rigorous contiguity is not a panacea, 14 states condone annexation of noncontiguous territory under special circumstances, such as city-owned land with no residents—a reasonable strategy for protecting a city's interest in watershed land or a new industrial park.[23] What's more, Indiana, Kansas, Nevada, and North Carolina permit noncontiguous annexation when all property owners agree and the city can guarantee fire and police protection and other municipal services. When annexation statutes insist upon strict contiguity, even for nearby city-owned land, the typical result is a "flag annexation" similar to the 250-foot-wide, seven-tenths-mile-long corridor tying O'Hare Airport to Chicago, which annexed the site in 1956, prior to making it the city's main airport.[24] Los Angeles uses a longer and wider "shoestring"— 10 miles long and a half-mile wide—to keep the port at San Pedro within the city limits.

Dramatic annexations like these are not the norm. The 230,271 annexations counted in the United States between 1970 and 1998 added an average of only 0.13 square miles to an existing municipality.[25] More intriguing is the geographic concentration of activity. Seven states (Illinois, California, Florida, Texas, Georgia, North Carolina, and Alabama, in that order) account for over half the national total, and Illinois, with 30,830 annexations—an average of more than a thousand per year—registered more annexations than the 29 least active states combined. At the bottom of the list, Hawaii and the six New England states collectively reported only seven annexations, and New Jersey, the next highest, registered only 30. Illinois not withstanding, annexation was more common in recent decades in Sunbelt states, where in-migration underlies the rapid growth of small- and medium-size cities.

County boundaries are comparatively stable. Altering a state's primary subdivisions requires legislative approval and is not taken lightly. There are only three options: existing counties can split or merge, or boundaries between neighboring counties can be moved, perhaps with an exchange of territory. Change became inevitable after population growth and emerging cities made the original boundaries intolerably inefficient. John Long, editor of the *Atlas of Historical County Boundaries*, reckons that the average county in the country's older states experienced four and a half boundary changes, with some registering more than two dozen "boundary change events."[26] Onondaga County, New York, where I live, reflects a typical fission into smaller, more manageable units with county seats no farther than a half day away by horse or boat—that's why states settled after steamboats and railroads became common tend to have large counties.[27] Created in 1794 from parts of two other counties, Onondaga hived off three new counties in 1799, 1808, and 1816, respectively.[28] Its fourth and final boundary change occurred in 1829, when an exchange of land with its western neighbor simplified their common boundary along a small lake. The state registered only 12 boundary changes since 1900, the last in 1964, when a small island in the East River was transferred from Queens County to Bronx County.[29] Boundary adjustments in the country's younger states have been proportionately recent.

Rigid States

State boundaries are comparatively rigid, fixed at the time of admission to the union and largely immutable except for minor adaptations to avulsion. While it might make sense, both culturally and logistically, to have a North California and South California with separate capitals and governments like those of North Carolina and South Carolina, don't expect the rest of the country to award our most populous state two additional seats in the Senate. Although two Senators each for California and Wyoming, our least populous state, goes against the notion of one-person-one-vote, this inequality is firmly embedded in the Constitution, and thus certain to be upheld by the Supreme Court. Although the Constitution can be amended, reconfiguring state boundaries is a slippery slope, likely to create more losers than winners among the existing states, three-quarters of which would have to approve. Because a reconfiguration would most likely involve fusion as well as fission, don't look for support from New Hampshire and Vermont—sharing a page in the *Rand McNally Road Atlas* is bad enough, but losing their clout in the Senate as well as their disparate approaches to taxation and government services is unthinkable.

However unlikely, it's intriguing to speculate on a more rational configuration of state boundaries. A number of people have tried, with little acclaim and no lasting impact.[30] The most prominent was G. Etzel Pearcy,

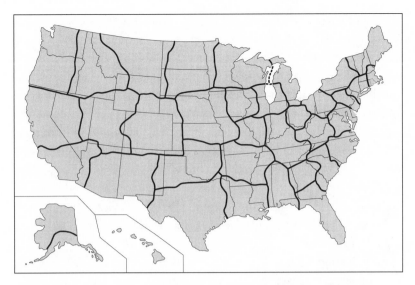

FIGURE 6.6. A small-scale generalization of G. Etzel Pearcy's "Thirty-Eight State U.S.A."

who retired as chief geographer at the State Department in 1969, and moved west to work for several years as a geography professor at Cal State, Los Angeles (not to be confused with UCLA). In 1973, Pearcy published *A Thirty-Eight State U.S.A.*, in which he redrew boundary lines and named the resulting provinces after a regionally significant physical or cultural feature.[31] Hawaii, the only state to survive intact, was renamed Kilauea, after the fitfully active volcano that dominates the Big Island, while a somewhat reduced version of Texas was dubbed Alamo, and the province centered on Boston became Plymouth. The only names to survive (sort of) were Carolina and Dakota, whose borders reflect vague (at best) amalgamations of their northern and southern antecedents.

Pearcy based his reconfiguration (fig. 6.6) on efficiency and cultural homogeneity, not visual prominence and ease of mapping. Slavish alignments along meridians and parallels were out, as were river boundaries, which denied advances in bridge-building, diminished the importance of watersheds as ecological units, and increased the risk of tedious litigation following a sudden or not-so-sudden change in the river's course—or cartographic oddities like crescent-shaped enclaves of Louisiana and Mississippi left behind within each other's borders. Routing boundaries through areas of relatively low population density made it easier to favor compact shapes, minimize disparities in size, and keep metropolitan areas intact. Each state contained at least one prominent regional center, which makes sense to geographers. Ridiculously small states like Rhode Island and Delaware were out, but so were huge, dispersed territories like Alaska and Texas. For Pearcy, 38 seemed an "optimal number"—an efficient compromise between too many states, "involving an overly heavy bureaucratic structure [which] unnecessarily adds to the cost of state administration," and too few, which might "overcharge the responsibilities of centralized control and weaken the effectiveness of administration."[32] *Time,* which framed Pearcy's map as a plan for reducing the overall cost of state government, quoted him as saying, "I know of several Governors we could do without."[33] No doubt a few governors were similarly leery about retired State Department geographers.

In preparing a generalized picture of Pearcy's 38-state makeover (fig. 6.6), I abruptly realized that he was concerned more with reducing the total number of states and affixing clever names than with delineating optimal boundaries. Short descriptions of each new state include a small map relating its borders to major cities and rivers, but the boundaries on these cartographic vignettes don't always match the corresponding lines on the double-page, centerfold map covering the entire country. The most egregious example is Erie, described as including Cleveland and Cincinnati as well as a substantial portion of Lake Erie shoreline (fig. 6.7, *left*) but left portless on Pearcy's national map (fig. 6.7, *right*). Little point, I suppose, in fine-tuning boundaries unlikely to be adopted, much less debated.

Instead of condemning our historic state boundaries, screenwriter and drama teacher Mark Stein celebrates them in *How the States Got Their Shapes.*[34] Many states started out on paper with shapes quite different from their current configuration. Maryland, for instance, might have had a northern border through present-day Philadelphia had Lord Baltimore promptly surveyed and settled his colonial grant in the 1630s. When Marylanders became aware of their entitlement, Pennsylvania had a well-established foothold on the Delaware River. Decades of legal wrangling in the English courts led to the compromise boundary known as the Mason-Dixon Line, which includes a north-south border with Delaware as well as the more famous east-west border with Pennsylvania. Look carefully at a map (fig. 6.8) and you'll notice that a short extension of the Maryland-Pennsylvania line eastward to the circular arc that gives Delaware and Pennsylvania the most distinctive interstate boundary in the country.

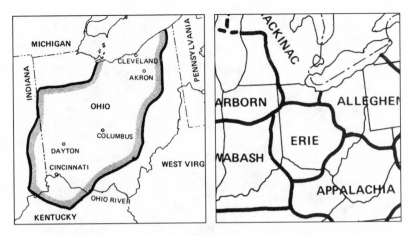

FIGURE 6.7. Disparity between Pearcy's individual map of Erie (*left*) and its treatment on his national map (*right*).

Delaware started out as a Dutch colony, but after the British displaced the Dutch in the 1670s, William Penn and the Duke of York tied the southeastern boundary of Pennsylvania to a "circle drawne at twelve miles distance from Newcastle"—easily declared on paper if you don't have to survey it.[35]

Although most maps depict Delaware's "Twelve-Mile Circle" as a perfectly circular arc, this cartographic oddity actually consists of parts of two circles with radii of 12.81 and 11.59 miles, centered respectively on points a little southeast and a little northeast of the arc's reputed center atop the courthouse in what's now New Castle.[36] It was supposed to be a single circle, but the early surveys didn't get it right—and probably couldn't. Because a functional boundary had already been marked, a comparatively precise 1892 survey compromised on a "compound circle," demarcated by 42 monuments a half mile apart.

This is not the only quirk in Delaware's perimeter. Truth be told, the state's western boundary with Maryland consists of three lines: a short arc with a radius of 12 miles and 108 feet actually centered on the courthouse, a line running due north about four miles to Maryland's northeast corner, and in the opposite direction a much longer portion tangent to the circle and trending just a bit eastward of due south.[37] This trio of connecting alignments account for the ever-so-subtle bend at the north end of the Delaware-Maryland border (fig. 6.8).

A third anomaly underscores the restrictive impact of state boundaries.

FIGURE 6.8. Delaware's perimeter includes a circular border with Pennsylvania, a slightly kinked western border with Maryland, and a maritime boundary along the New Jersey riverfront.

In most instances where a river divides two states, their boundary approximates the middle of the river, typically the deepest part of the main navigation channel—geographers call this the *thalweg*, a German word meaning "the way of the valley." Not so in the Delaware River. New Jersey was formed after Delaware, which retained ownership of the river bed within the Twelve-Mile Circle. New Jersey challenged this division several times, starting in 1877, but a 1935 Supreme Court ruling fixed the boundary within the circle along the mean low-water line, as observed in 1934. New Jersey can control activities on the riverbank, but Delaware's jurisdiction begins at the 1934 low-water line. As two small patches on the map (fig. 6.8) confirm, land created since then by filling in the river is part of Delaware even though it's attached to New Jersey. Outside the Twelve Mile Circle, the boundary follows a more typical path.

New Jersey went back to court in 2005. This time much more was at stake than fishing licenses and oyster beds. Two years earlier BP, the international energy conglomerate, had proposed a liquefied natural gas (LNG) terminal along the Jersey shoreline, about 10 miles north of New Castle and conveniently connected to regional gas distribution lines. Two or three times a week a huge tanker from Trinidad would dock at the end of a 2,000 foot pier and unload enough low-cost fuel to meet residential demand in Philadelphia, several Pennsylvania counties, and all of Delaware and New Jersey.[38] Fearing environmental damage—LNG tankers make great terrorist targets—Delaware officials refused to issue a permit. In 2007, a Special Master appointed by the Supreme Court determined that Delaware was within its rights, and in 2008 the justices agreed that the state's right to control its river trumped New Jersey's eagerness to develop its shoreline.[39]

New Jersey had fared better a decade earlier, when it challenged New York State's sovereignty over Ellis Island, where 12 million immigrants entered the country between 1892 and 1954. Like most litigation, this one involved both pride and money—the island is a unique part of our national heritage, and visitors to its museum pay sales tax on food and souvenirs. The dispute was rooted in an awkward double boundary established in 1834.[40] In brief, New Jersey received half the land under the Hudson River, but on the surface its sovereignty stopped at the river's edge, including docks and filled land extending beyond the natural shoreline. Paradoxically, the submerged-land boundary runs between Staten Island and the rest of New York City, while the surface boundary hugs the western shoreline of Arthur Kill (fig. 6.9, *left*). New York received all islands in the harbor, including Ellis Island, a mere three acres at the time. When Congress loosened restrictions on immigration in the late nineteenth century, the government enlarged the island and built a "reception center" for processing new arrivals. New Jersey argued that the filled land was part of the Garden State, and in 1998 the Supreme Court agreed.[41] Lacking an 1834 map, the court took the island's original perimeter from an 1857 Coast Survey chart.[42] New York still collects sales tax from concessions within the donut hole (fig. 6.9, *right*), while New Jersey taxes purchases elsewhere on the site.

Bureaucratic Bordering

Two final examples illustrate the diversity of borders accompanying the radical expansion of government in the twentieth century. This ramped-up range of roles demanded a more geographically nuanced form of public administration, impossible without more finely textured administrative boundaries. Where these ad hoc borders often conferred benefits or imposed penalties, they drew proponents and detractors.

FIGURE 6.9. Under the Compact of 1834 New Jersey and New York agreed to a double boundary that differentiates jurisdiction above the water line from jurisdiction over submerged lands (*left*). In 1998 the Supreme Court limited New York's jurisdiction on Ellis Island to the island's original extent in 1834 (*right*).

Negotiations typically occurred in legislative committees or behind-the-scenes meetings with lobbyists. One result was a vast expansion in the number of Metropolitan Statistical Areas (MSAs) in the 1970s and 1980s, when hundreds of whole counties were relabeled "metropolitan" to make them eligible for increased public-housing and mass-transit subsidies. Small cities within these newly designated MSAs feared loss of housing programs reserved for areas "rural in character," but the Rural Housing Service, the Department of Agriculture agency that determines eligibility, "grandfathered" concentrations of more than 10,000 residents within MSAs as long as their populations didn't exceed 25,000, and they remained essentially "rural." Responding to complaints about fairness, the Government Accountability Office (GAO) concluded that "delineating eligible areas from ineligible areas . . . is time-consuming, requires judgment, and can be problematic," and that "apparently similar rural areas received different designations."[43] The GAO recommended dropping both the grandfather clause and the MSA exclusion and basing "rural" on measurements of population density and commuting behavior—too much tract housing and too many ties to the big city, and you're out.

Belated recognition of Native American sovereignty led to another kind of administrative construction, the land trust within which property purchased by a tribal government outside its reservation boundary is exempt from local taxes and regulations. Indian nations enjoy substantial autonomy within their sovereign territory but unless placed in a land trust administered by the Bureau of Indian Affairs (BIA), holdings elsewhere are little different from land owned by you or me. Moreover, not all tribal purchases qualify. In Central New York, for instance, the BIA required a comprehensive environmental impact statement before giving trust protection to the Oneida Indian Nation—but only for parcels near the original reservation or the tribe's new resort and casino.[44] A more questionable use of land-trust privileges occurred in Utah, where an outdoor advertising firm from California helped the Shivwits Band of Paiute Indians purchase land near Interstate 15 within the city limits of St. George.[45] The tribe put the parcels in a BIA land trust and then leased it to the firm, which began to erect billboards. City officials protested this apparent violation of the Highway Beautification Act, but the U.S. Court of Appeals ruled that the tribe was within its rights—the city might have won, but rampant development within its boundary undermined St. George's argument against visual pollution. As with other creatures of administrative bordering, the courts can override or reinforce BIA land trusts.

CHAPTER

SEVEN

....

CONTORTED
BOUNDARIES,
WASTED
VOTES

A reviewer once complained that my book on redrawing congressional districts offered "no clear point of view."[1] While the review was generally favorable, the remark stung because I thought I had outlined a workable solution, radical perhaps yet surely constitutional. Apparently, I'd put a bit too much on the table—though logical and vigorous, my argument was packaged with an intentionally broad assessment of political cartography that included the historical context, international comparisons, a generous dose of anecdotes too engaging or insightful to ignore, and some technical details essential to understanding why some of the districts drawn up after the 1990 Census were more grotesque than the prototypical 1812 Gerrymander. With far fewer distractions, this chapter argues emphatically that America's map-obsessed approach to electoral redistricting distorts not only the geography of voting districts but also the spirit of representative democracy.

A secondary point—and the only reason for including voting districts in this exploration of prohibitive cartography—is that partisan gerrymandering produces overwhelmingly Democratic or Republican districts, which make many voters feel their ballots are wasted. While it's obvious that not everyone's preferred candidate can win, America suffers a dearth of what political scientists call competitive districts, in which party affiliation is not the key to victory. As I'll show, an alternative electoral cartography uses stable, geographically meaningful boundaries to ensure wider recognition of citizens' concerns as well as increase voter turnout. But first we need to look at how maps manipulated by partisan politicians waste votes and undermine the Supreme Court's one-person-one-vote doctrine.

Gerrymanders and Bushmanders

To understand how gerrymandering works, you must look at the numbers as closely as you look at the map. Some arithmetic is unavoidable, but I'll keep it simple with an example for a fictitious state with 3 million people split evenly among three congressional districts. To keep the geography simple, my hypothetical state consists of 54 counties, all square and identical in size. Each of its three urban counties has 150,000 residents, and each of the remaining 51 rural counties has 50,000 residents. To keep the politics and demographics simple, I assume a two-party system in which everyone is either a Petulant or a Stalwart, party loyalty is ingrained, and everyone votes. To make this example instructive, I give the Stalwarts a slight edge (27,000 to 23,000) in the rural counties but let the Petulants enjoy an overwhelming majority (120,000 to 30,000) in the urban counties, each hosting one of the state's three universities. Do the math, and you'll see that in a statewide contest, the Petulants enjoy a 1,533,000 to 1,467,000 advantage:

Petulants:

$$(3 \times 120,000) + (51 \times 23,000) = 360,000 + 1,173,000 = 1,533,000$$

Stalwarts:

$$(3 \times 30,000) + (51 \times 27,000) = 90,000 + 1,377,000 = 1,467,000$$

With this margin, the Petulant candidate will have no difficulty winning the governorship, and in a fair society, party officials should be perfectly content to capture only two of the state's three congressional seats.

Party officials, not surprisingly, have a notion of fairness that eagerly

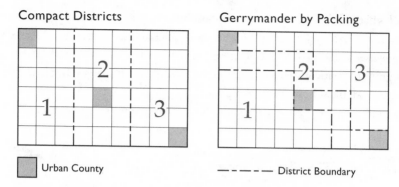

FIGURE 7.1. While compact districts each with an urban center (*left*) favor the party with a slight statewide majority, packing the majority party's voters into a long, thin highly urban district (*right*) gives the minority party a decisive advantage in two of the state's three districts.

exploits a marginal advantage as fully as tradition and the law allow. Because redistricting is traditionally controlled by the majority party, which enjoys enormous freedom to serve its own interests as long as districts are essentially equal in population and there's no apparent racial discrimination, Petulants can argue vigorously for the three compact districts in the left half of figure 7.1. Each has a single urban center, a uniform shape, and a distribution of voters marginally advantageous to the Petulants:

Petulants:

$$(1 \times 120,000) + (17 \times 23,000) = 120,000 + 391,000 = 511,000$$

Stalwarts:

$$(1 \times 30,000) + (17 \times 27,000) = 30,000 + 459,000 = 489,000$$

If the customary attack ads, lawn signs, and get-out-the-vote electioneering work, Petulant leaders will no doubt applaud the wisdom of winner-take-all voting, which assures them all three seats.

But suppose there's an upset, precipitated perhaps by a racial gaff, a sex scandal, or an untimely DWI arrest? And suppose this upset occurs shortly after the decennial population census so that the Stalwarts also win the right to reconfigure voting districts? If so, don't be surprised when their spokespeople argue that uncomplicated borders are less relevant than coherent constituencies—in this case, the single distinctly urban district and the two comparatively rural districts in the right half of figure 7.1. Packing Petulant voters into a long, thin urban district insures a landslide victory for whichever Petulant incumbent survives a contentious post-remap primary:

Petulants:

$$(3 \times 120{,}000) + (11 \times 23{,}000) = 360{,}000 + 253{,}000 = 613{,}000$$

Stalwarts:

$$(3 \times 30{,}000) + (11 \times 27{,}000) = 90{,}000 + 297{,}000 = 387{,}000$$

Despite this dramatic trouncing in the tri-city district, Stalwart leaders will privately if not publicly praise the prowess of the political cartographers responsible for their decisive edge in the other two districts:

Petulants:

$$(0 \times 120{,}000) + (20 \times 23{,}000) = 0 + 460{,}000 = 460{,}000$$

Stalwarts:

$$(0 \times 30{,}000) + (20 \times 27{,}000) = 0 + 540{,}000 = 540{,}000$$

While the previous configuration wasted Stalwart votes by giving the Petulants all three seats, the new plan wastes Petulant votes by awarding the Stalwarts not one, but two seats.

Packing opposition support into odd-looking districts is the quintessential form of gerrymandering, as pioneered in 1812 by the Jeffersonian Republicans who controlled the Massachusetts legislature. When population shifts revealed by the 1810 Census called for remapping the state's senatorial districts, the Republicans hoped some inspired political cartography could win them a few more seats. Among other ploys, they split Essex County into a solid Federalist stronghold in the center and southeast and a long, thin, and marginally Republican district along the northern,

FIGURE 7.2. Artist Elkanah Tisdale crafted the famous gerrymander cartoon by embellishing the map of senate districts devised for Essex County by the Massachusetts legislature. His caricature was first published in the *Boston Gazette* for March 26, 1812.

western, and southwestern fringes. The governor at the time was Elbridge Gerry, who pronounced his name with a hard *g*, as in Gary, Indiana. Gerry cringed at the awkward shape but was reluctant to veto his party's redistricting bill.[2] When a reporter for a Federalist newspaper pointed out a faint resemblance to a salamander, his editor exclaimed, "Salamander! Call it a Gerrymander!" This remark inspired artist-cartoonist Elkanah Tisdale to create the well-known gerrymander cartoon (fig. 7.2), which irrevocably linked the governor's name (typically mispronounced with a soft *j*, as in Jerry) with cartographic manipulation.[3] Historians look more kindly on Gerry, who in 1789 proposed what became the Library of Congress and later served as vice president under James Madison.

A new political monster emerged in the 1990s, when pressure from the Justice Department forced political cartographers in several states to create new "minority-majority" congressional districts, so called because they packed nonwhites into districts in which African Americans or Hispanics constituted markedly more than a majority of the voting-age population, and thus gained sufficient strength to elect one of their own to the House of Representatives.[4] As an expansive interpretation of the Voting Rights Act, the strategy benefited Republicans because African Americans

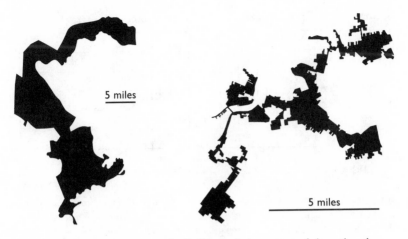

FIGURE 7.3. The original gerrymander (*left*) had a far less contorted shape than the New York's Bullwinkle District (*right*) and other post-1990 Bushmanders.

and Hispanics tend to vote Democratic—packing them into minority-majority districts left adjacent districts with fewer Democrats, and thus more likely to elect Republicans. Even so, the Democrats who controlled redistricting in the states affected were determined to protect their own incumbents wherever possible. With the aid of powerful geographic information systems and census data for city blocks and other small areas, they tweaked the boundaries as much as they could without aggravating the Justice Department's map police. The resulting shapes were markedly more complex than the 1812 gerrymander (fig. 7.3, *left*), and inspired names like "Bullwinkle District" for a Hispanic constituency in New York City with antler-like prongs that recalled the drawling moose featured in Saturday morning TV cartoons (fig. 7.3, *right*). Because George Herbert Walker Bush was in the White House at the time, I dubbed these new shapes Bushmanders, and irritated Republican friends when I titled my book *Bushmanders and Bullwinkles*—an apt title insofar as George H. W., like Elbridge Gerry, was more a bystander than an instigator.

Packing is not the only ploy for political cartographers seeking partisan advantage. As figure 7.4 demonstrates for a different configuration of my hypothetical state's urban counties, an equally potent variant called cracking can dismantle a cohesive metropolitan district. If the Petulants in this new geography held a weaker advantage in urban counties and depended on the three-county urban district in the center of the left-hand map to win just one of the state's three seats, cracking their metropolitan strong-

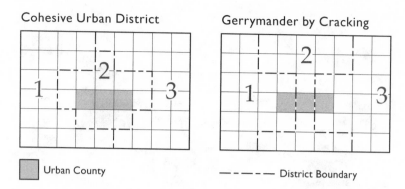

Cohesive Urban District | Gerrymander by Cracking

■ Urban County   — — — — — District Boundary

FIGURE 7.4. Cracking is the converse of packing. Depending on the numbers, cracking can dilute the strength of an opposing party with a marginal majority in a cluster of urban counties—or redistribute its strength advantageously into neighboring districts.

hold could give the Stalwarts all three seats. But with the comparatively advantageous numbers used earlier, self-inflicted cracking would clearly benefit the Petulants, who could win all three seats by shifting some of their 120,000-to-30,000 urban-county advantage into neighboring districts. As shrewd politicians are well aware, the map's power depends on both boundaries and numbers.

Because awkward boundaries are not the only way to make the numbers favor a particular party, gerrymanders need not be as visually blatant as the meandering urban district in figure 7.1—modern redistricting software can use shape measurements to seek out flagrantly partisan gerrymanders that look geometrically respectable.[5] Insofar as the goal is to waste votes for the other party's candidates, the left-hand map, which denies the Stalwarts a single victory, is as much a gerrymander as the right-hand map, which gives them two seats. The configuration of urban counties and our conditioned respect for compactness just makes the boundaries in left-hand map look more natural, especially if we don't highlight the urban counties, two of which occupy a corner in their districts. Similarly, the boundaries in figure 7.4, which shifts the focus from packing to cracking, demonstrate that whichever party controls redistricting should be able to win one more seat than its overall statewide strength suggests. By contrast, a more equitable plan, with a likely outcome consistent with the Petulants' small but decisive statewide majority, would give the Stalwarts an advantage in a single district with 20 rural counties and the Petulants an edge in two districts, one with one urban and 17 rural counties and the other with two

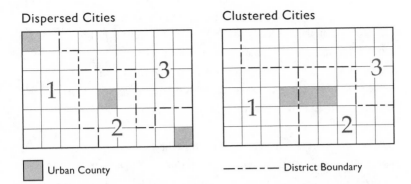

FIGURE 7.5. Plausible reconfigurations of the districts in figures 7.1 and 7.4 to create an arguably more equitable outcome that wastes fewer votes by giving the Stalwarts one seat and the Petulants two. Other reconfigurations with identical allocations of urban and rural counties will yield similar electoral results.

urban and 14 rural counties—if you don't believe me, do the math. And as figure 7.5 demonstrates, solutions that waste fewer votes statewide exist for both dispersed and clustered arrangements of urban counties.

Shaping Boundaries or Shaping Ballots

Many, perhaps hundreds, of arguably equitable redistricting plans exist, which raises the thorny questions of which to adopt and who decides. Without radical reform, there's a strong chance that boundaries will be tweaked, if not flagrantly manipulated, to favor the majority party, a clique of privileged incumbents, or both. Although redistricting could be turned over to an "independent" commission, most schemes let politicians influence the board's composition or veto its decision. Without broad public participation, there's a strong chance of a highly partisan map and wasted votes. By contrast, asking 3 million people to configure the boundaries seems ludicrously impractical—unless we replace the conventional map and its readily manipulated borders with a voting system that, in essence, lets voters redistrict themselves.

This concept is hardly new. It's used in Australia and several European countries, most notably Ireland, as well as a handful of U.S. cities, including Amarillo, Texas, and Peoria, Illinois. Dubbed "full representation" by advocates, the approach has variations with often-puzzling labels like "choice voting," "cumulative voting," "limited voting," "preference voting," and "proportional representation."[6] While the "full" in full representation is somewhat misleading insofar as some people's votes are bound to prove useless, at least for the more extremist parts of the political landscape, in a state with two or more congressional representatives, full representation maximizes the likelihood that the average voter will influence the election of *at least one* representative.

A key element of full representation is the multimember district. Our hypothetical state with three single-member districts thus becomes a three-member district, in which each party fields a slate of three candidates, joined most certainly by an assortment of independent aspirants. Although all candidates would run statewide, full representation rejects the traditional notion of "first past the post," whereby 51 percent of the voters could fill all three seats. The most defensible variant, known as choice voting in Cambridge, Massachusetts, lets each voter rank all candidates deemed acceptable—1 for the most preferred, 2 for the second most preferred, and so forth. If each of the two parties in our hypothetical state fields three candidates, a Stalwart voter could designate a fourth-, a fifth-, and even a sixth-place choice among the three Petulant nominees, while a Petulant voter could specify similar lower-rank preferences for moderately acceptable Stalwart candidates. The system is most effective when voters are not wedded to party endorsements and when election laws encourage independent candidates to circulate nominating petitions.

Choice voting's other name, "the single transferable vote," hints at how votes are counted and winners declared. To gain a seat in a multimember district, a candidate must meet a "quota" calculated by first dividing the number of ballots cast by the number of seats plus one, and then adding one to the result.[7] The quota, which represents the minimum number of votes needed to win a seat, also determines the number of excess votes, which are transferred to voters' next-in-line choices. For example, if all 3 million residents of our hypothetical state go to the polls, the resulting quota of 750,001 guarantees a seat to any candidate with this many (or more) first-place votes.[8] If a candidate garners 800,000 first-place votes, the system then redistributes the 49,999 excess votes on a pro rata basis to the other candidates. Because only 750,001 votes are needed to win a seat, the remaining rankings on each of these 800,000 ballots is treated as 6.625 percent (49,999 ÷ 800,000) of a vote for their second-ranked choice. Had the most preferred candidate received 1,500,002 votes—twice as many as needed—the 750,001 excess votes would be transferred as 1,500,002 fifty-percent (750,001 ÷ 1,500,002) votes. Because excess votes are not wasted, there's no risk in giving a highly popular candidate your first-place rank.

Tabulation begins by counting the number of first-place votes for all candidates. Any who exceed the quota are awarded seats, and their excess votes are distributed to the remaining candidates. Another round of counting determines which, if any, of the remaining contestants meet the quota. Any who do are awarded seats and their excess votes are redistributed to candidates still in the race. The focus then shifts to bottom of the list by eliminating the weakest remaining aspirant, redistributing his or her votes, checking whether anyone else now meets the quota, and repeating the process until all remaining seats are filled.[9] Before computer tabulation was introduced in 1997, Cambridge residents had to wait as long as a week to learn the outcome of city council and school committee elections. Unofficial results are now available the same day, in ample time for late evening TV news.[10]

Cambridge voters, who appreciate their complicated ballot, have rejected numerous attempts to restore winner-take-all elections, more easily manipulated by partisan politicians. Voters also understand the fairness in printing as many different ballots as there are candidates, so that no contender enjoys a consistently preferential position. While many jurisdictions let the leading party in the last election put its candidates at the top, Cambridge rotates an alphabetical list so that each name appears first on an equal number of ballots. Candidate Adams thus enjoys no greater advantage than Candidate Zelinsky.

Because Cambridge voters have ready access to the same newspapers and television stations, treating the entire city as a single multimember district is hardly unreasonable even though there are nine seats on the city council. By contrast, subdivision of some sort is essential for huge states like California, New York, and Texas, with multiple media markets and many seats to fill in Congress and the state legislature, and partitioning medium-size states like Colorado and Wisconsin makes sense as well. FairVote, an advocacy group also known as the Center for Voting and Democracy, favors dividing the country's more populous states into "superdistricts" each with three to five seats in Congress.[11] This range is large enough to let a substantial minority group choose one, or perhaps two, of the district's congressional delegation yet small enough for efficient campaigning and informative media coverage. Opportunities for female and third-party candidates should be noticeably improved as well.

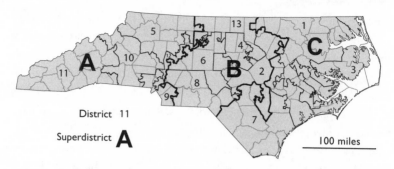

Superdistrict **A**

District 11

Superdistrict **A**

100 miles

FIGURE 7.6. Superdistricts created by FairVote around 2005 by combining three to five of North Carolina's 13 congressional districts.

To illustrate what superdistricts might look like, FairVote split all states with six or more seats in the House of Representatives into two or more superdistricts by combining congressional districts used to elect the 109th Congress. Although this strategy is admittedly flawed because existing single-member districts often reflect partisan gerrymandering, the maps illustrate geographic principles useful in crafting more meaningful super-districts. For example, FairVote's partition of North Carolina, with 13 House seats, into three superdistricts (fig. 7.6) acknowledges salient geographic differences between a comparatively mountainous western region prominent in recreation and tourism, a central Piedmont region noted for research and manufacturing, and an eastern coastal region oriented to agriculture, fishing, and marine recreation.

The geography seems right, but what about the politics? A cursory political analysis suggests that each superdistrict would have one competitive seat, not stacked in favor of either a Democrat or a Republican, and that African American voters—if they chose—could easily elect one of their own in the central and eastern districts, which have five and three seats, respectively. Although FairVote's three superdistricts would probably not change the number of African Americans, Democrats, and Republicans the state sends to Congress, the perception of wasted votes would be much lower than for the actual post-2000 remap, with two districts deliberately configured to favor black candidates (1 and 12), one additional district gerrymandered to favor a Democrat, and six districts configured to elect Republicans.[12] Very few voters would be disappointed by having no say in their superdistrict's congressional delegation.

A similar amalgamation of Pennsylvania's existing 19 districts into five

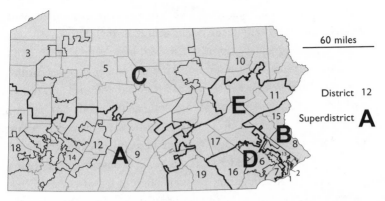

FIGURE 7.7. Superdistricts created by FairVote around 2005 by combining three to five of Pennsylvania 19 congressional districts.

superdistricts (fig. 7.7) yields less contorted boundaries in most places and recognizes important regional constituencies in mountainous northern and southwestern parts of the state as well as in the generally less rugged region between Scranton and York, in the lowland counties in the southeast, and in Philadelphia and its northern suburbs. As in North Carolina, a competitive "swing seat" in each superdistrict helps minimize the number of wasted votes.[13] Free to craft new superdistricts from scratch, rather than merely agglomerate gerrymandered single-member districts, a competent geographer or political scientist could much better approximate media markets and accommodate significant economic, cultural, and recreational differences.

Arguments against superdistricts and choice voting have focused on the computational demands of complex ballots and the possible election of third-party or fringe candidates.[14] Neither objection is convincing. Electronic computers and the recent revamping of voting machines have demolished the first obstacle, and recurrent factional squabbles suggest that the two-party system is neither the most natural nor the most efficient way of ensuring broad representation and resolving disputes. Our republic has not only survived but benefited from third parties, and the experiences of numerous Western democracies confirm that vigorous public discourse in multiparty general elections is hardly destabilizing. But to give superdistricts a chance, Congress must first repeal an obscure 1967 law banning multimember congressional districts.[15] And before this can happen, the country must become more comfortable with multimember districts and choice voting at the municipal level, where it's easier to experiment with letting citizens redistrict city councils and school boards.[16]

Let me suggest a further improvement, even more radical than multi-member superdistricts. Weighting each House member's vote according to population could greatly reduce the need to split counties and redraw boundaries every 10 years. Instead of triggering another round of partisan gerrymandering, the population census could be used instead to adjust upward or downward the weighted votes of members of Congress. Think of proxy voting at a stockholders meeting: each House member votes a number of shares proportional to the population of his or her district— for multimember constituencies, district population is first divided by the number of seats. A superdistrict growing more rapidly than the country as a whole would gain shares, but its boundaries need not change. Similarly, a superdistrict growing less rapidly would lose shares, but its boundaries could remain intact. While a roll-call vote weighted by population might seem unwieldy, if not strangely un-American, it's hardly less fair or logical than Congress's committee structure, which privileges seniority and party affiliation. Because these traditions would remain intact, a persuasive and gifted House member with a 90 percent vote need be no less influential than a colleague with a 110 percent vote.

Boundaries might need to be redrawn occasionally to reflect population shifts as well as profound changes in regional coherence. In most cases, though, when a state gained or lost a seat because of post-Census reapportionment, one of its superdistricts would absorb the gain or loss. Because a superdistrict's total vote must be proportional to its population, representatives from a district gaining a seat would typically see their weighted votes decline while each representative from a district losing a seat would enjoy a larger weighted vote. Of course, persistent, multi-decade growth or decline might require splitting or merging superdistricts to keep the number of seats between three and five, and a few of the smaller states might need a single two-member district. Apparent anomalies are unavoidable. But by combining population equality with meaningful geography, weighted voting in multimember superdistricts can put an end to hypocritical election-district maps that say "No!" by wasting votes and discouraging participation.

CHAPTER EIGHT

....

REDLINING AND GREENLINING

Much of the map's power—if indeed *power* is the right word—lies in its predictive prowess, assumed or real. As examined in the preceding chapter, maps of gerrymandered single-member election districts can discourage voters from going to the polls—after all, why bother to vote when party registration tallies and winner-take-all elections suggest that one's preferred candidate hasn't a chance? Maps that prophesy also affect other significant decisions in civil society, notably the choice of where to live. Because homebuyers usually finance their purchase with a mortgage, the cartographic crystal balls of wary financial institutions can become self-fulfilling prophesies when workers with good credit histories who want to buy homes in African American or mixed-race neighborhoods have difficulty getting mortgages or insurance. Dubbed redlining after the map symbol for financially risky areas, delineations too easily based on race were outlawed by fair-housing legislation in the late 1960s. Even so, redlining persists in other contexts, often with ZIP codes as a cartographic surrogate. This chapter explores the impact of redlining as a form of prohibitive cartography and contrasts it with greenlining, a no less rhetorical place-based approach to economic "empowerment."

Redlining

Redlining gained prominence as a pejorative in the late 1960s, when community activists in Chicago condemned local saving and loan associations for outlining no-loan areas in red.[1] After the race riots of 1967 highlighted the difficulty of insuring property in predominantly black neighborhoods, President Lyndon Johnson gave the term wider prominence by asking insurers to "end the practice of 'red-lining' neighborhoods and eliminate other restrictive activities."[2] Although federal and state legislation reined in the more blatantly racial forms of place-based discrimination in the mortgage and insurance industries, a few social scientists not only blamed redlining for much of the postwar urban malaise but pointed an accusatory finger at a long-forgotten New Deal agency, the Home Owners Loan Corporation (HOLC), and its bureaucratic kin, the Federal Housing Administration (FHA) and the Home Loan Bank Board. An early critic was urban housing scholar Charles Abrams, who in 1954 held the agencies responsible for promoting racial segregation by giving private lenders a system of real property appraisal that "read like a chapter from Hitler's Nuremberg Laws."[3]

Created in 1933 as an emergency measure to counter an ominous rise in foreclosures, the HOLC energetically refinanced home mortgages at lower interest rates for longer terms. A godsend for faltering banks and hard-pressed homeowners alike, the HOLC refinanced over a million mortgages, 80 percent of them successfully. That a fifth of its borrowers ultimately succumbed to foreclosure attests only to its willingness to take risks—without HOLC refinancing all of them would have lost their homes.[4] Even so, the agency was wary of deteriorating structures in declining neighborhoods and launched the City Survey Program to assess risk at the neighborhood level for all cities with more than 40,000 residents.[5] Between 1935 and 1940 more than 230 cities were surveyed, some more than once. Field agents canvassed residential areas block by block, looking for "favorable" and "detrimental influences," describing the age and state of repair of buildings, estimating the "negro" and "foreign-born" proportions of the population, and noting the race or ethnicity of any "infiltration" as well as the presence of "relief families."[6] Assessors also talked with real estate agents, bankers, and local officials, and grouped a city's blocks into a small number of contiguous, comparatively homogeneous areas—in Philadelphia, for instance, their first and second maps delineated only 29 and

60 neighborhoods, respectively. Not bound by wards, census tracts, postal zones, or other preexisting units, field agents focused on identifying and describing coherent neighborhoods. When the agency dissolved in 1951, its Residential Security Maps and related documents—more than 150 boxes altogether—went to the National Archives, where they were ignored for a quarter century.[7]

Chance discovery of the maps in the late 1970s by urban historian Kenneth Jackson led to charges that the HOLC had "initiated the practice of 'red lining' [by] devis[ing] a rating system that undervalued neighborhoods that were dense, mixed, or aging."[8] Grievances escalated as the story was told and retold. In *American Apartheid*, sociologists Douglas Massey and Nancy Denton argued that because "banks adopted the HOLC's procedures (and prejudices) in constructing their own maps, [the agency] not only channeled federal funds away from black neighborhoods but was also responsible for a much larger and more significant disinvestment in black areas by private institutions."[9] According to journalist Buzz Bissinger, whose *Prayer for the City* chronicles Philadelphia's struggle for economic survival in the late twentieth century, "HOLC appraisers damned the city [while] delight[ing] in the suburbs that ringed it, not simply because of the amount of undeveloped land there, but also because of the ethnic and racial purity."[10] With similar disgust, political scientist Douglas Rae, who critiqued the HOLC maps for New Haven, Connecticut, attacked "the evaluation scheme [for having] embedded within it a whole courtroom mob of 'hanging judges' for the urban neighborhoods."[11]

The maps were based on a four-level rating scheme that combined traffic-light colors with the letter grades familiar to all college students. Topping the scale were the A neighborhoods, colored green on the maps and described concisely in the legend as "best." B neighborhoods, highlighted in blue and labeled "still desirable," were followed by C areas, tinted yellow for "definitely declining." At the bottom were the D neighborhoods, portrayed in red and deemed "hazardous." Ratings were also represented numerically, as "Grade 1" for the best neighborhoods and "Grade 4" for the worst.[12]

Far more inflammatory were the accompanying evaluation sheets, which reflected the blatant biases of some field consultants. For example, characterizations accompanying the maps for Philadelphia describe one grade A neighborhood as "Good transport but no sewers; Property owned by population of German descent making section more desirable" while

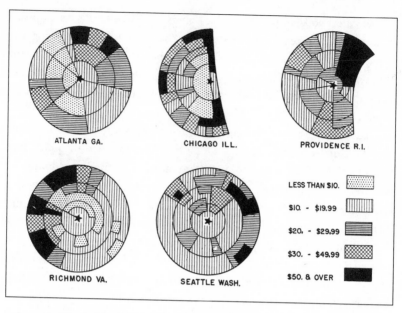

FIGURE 8.1. Homer Hoyt's "theoretical pattern" of average monthly rent of dwelling units in five American cities.

portraying a grade D area as "Close to business, heavy obsolescence; Concentration of foreigners and some Negro, some factories, predominance of lower class Jewish/Polish, Lithuanian, Slavs."[13] Although ethnicity and race were not the only perceived assets and deficiencies recorded, notes like "Negro encroachment threatening" for a transitional area in Pittsburgh and "not uncommon to find Negro houses adjacent to much better units" for a high risk neighborhood in the same city attest to institutionalized bias in the real estate and finance communities.[14]

Underlying HOLC neighborhood ratings was a theory of urban ecology devised by sociologist Robert Park and his colleagues at the University of Chicago in the 1920s and 1930s. Park conceptualized the city as a dynamic mosaic of sectors and concentric rings differing in land use and housing quality as well as in income, race, and ethnicity. Residential areas followed a life cycle of new construction, maturation, and blight, as comparatively affluent homeowners moved farther outward, passing the neighborhood along to less prosperous successors.[15] As illustrated for five cities in figure 8.1, sectors and rings frame a generalized pattern of average monthly household rent, a surrogate for neighborhood quality. I excerpted these examples from *The Structure and Growth of Residential Neighbor-*

hoods in American Cities, written by Homer Hoyt, a student of Park and an influential economic advisor to the FHA, which published his book in 1939.[16] Derived from detailed maps based on block-by-block surveys, the diagrams summarized the importance of sectors in predicting neighborhood quality. According to Hoyt, "the high rent neighborhoods of a city do not skip about at random in the process of movement—they follow a definite path in one or more sectors of the city."[17] Moreover, "if one sector of a city first develops as a low-rent residential area, it will tend to retain that characteristic for long distances as the sector is extended through process of the city's growth." Detailed mapping, he argued, was the key to forecasting risk to lenders and insurers.

Particularly intriguing to critics of New Deal housing policy is Hoyt's use of overlay analysis, which anticipated the electronic geographic information system (GIS) with an example based on Richmond, Virginia. Figure 8.2, a much reduced version of his original illustration, describes the superposition of four factors, identified by figure numbers at the upper right and tabs at the bottom.[18] In the background a map labeled "RENT" uses a lightly stippled pattern to highlight blocks with an average monthly rent less than 15 dollars. Glued to the paper along its left edge is a thin transparent sheet ("CONDITION") printed with horizontal lines to point out blocks where 25 percent of all structures "need major repairs or [are] unfit for use." Similarly attached are two additional overlay maps, one with diagonal lines sloping upward to the right ("AGE") to mark blocks where more than 75 percent of buildings are older than 35 years, and another with downward-sloping diagonal lines ("RACE") to locate areas where more than half the residents are nonwhite. Capping this sandwich of overlays is a summary map ("COMBINATION") on which solid black denotes the coincidence of "all 4 factors."

Although alignment of the overlays is less than perfect, it's clear that comparatively few blocks exhibit all four deficiencies. Hoyt noted enthusiastically that none of the Richmond's worst blocks were in its high rent areas while threatened areas were mostly near the city's center and toward the south and southeast. Systematic risk analysis was useful in all cities, he maintained, and "easily flexible" insofar as analysts with local knowledge could not examine a variety of factors relevant to the city in question but also experiment with thresholds.[19] In addition to map-based analysis of risk within a city, his method considered the local economy's long-term prospects, including "cyclical fluctuations in employment and diversifica-

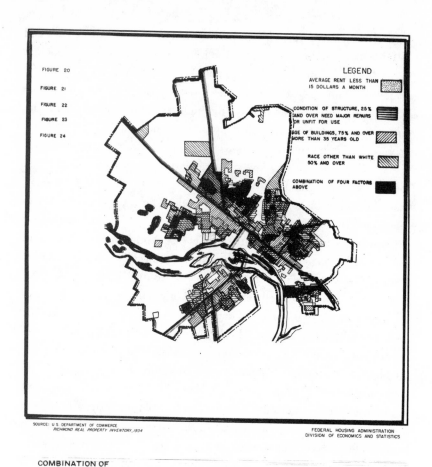

SOURCE: U.S. DEPARTMENT OF COMMERCE
RICHMOND REAL PROPERTY INVENTORY, 1934

FEDERAL HOUSING ADMINISTRATION
DIVISION OF ECONOMICS AND STATISTICS

COMBINATION OF
ALL 4 FACTORS   RACE   AGE   CONDITION   RENT

FIGURE 8.2. Homer Hoyt's graphic overlay of four transparent maps on a paper map of average rent describe the "coincidence of factors indicative of poor housing" for Richmond, Virginia, in 1934.

tion of employment [of] vital importance in evaluating mortgage risk."[20] At the national scale, redlining, or whatever one calls it, had regional as well as local impacts.

While local effects of redlining were no doubt painful, the HOLC's impact seems overstated. Although it's become fashionable to rail against the agency and its maps, their inflammatory visual rhetoric apparently had little direct effect on Depression-era home loans. Amy Hillier, a social scientist who used GIS to examine HOLC practices in Philadelphia, discovered that the agency made three-fifths of its loans in red areas and another fifth in yellow neighborhoods.[21] Although it stopped writing mortgages in 1936, before the maps were available, the HOLC treated them as confidential documents, available only to government officials.[22] Fearing misinterpretation, it produced no more than 60 copies of each sheet and required that all maps be returned to its statistical unit when no longer needed. Any correlation between red areas and mortgage discrimination most certainly reflects earlier biases—while redlining did not enter the political lexicon until the late 1960s, the practice was common among private lenders in the 1920s, and probably much earlier. The real culprit, according to Hillier, is the FHA, which promoted systematic place-based appraisal methods in its *Underwriting Manual,* first published in 1935 and widely distributed to bankers and real estate firms. "Whether or not it used maps with red lines," she concluded, the "FHA did more to institutionalize redlining than any other agency by categorizing mortgages according to their risk level and encouraging private lenders to do the same."[23]

Another form of redlining persisted long after Congress outlawed discriminatory lending. ZIP codes, introduced by the U.S. Postal Service in 1963 to expedite the delivery of mail, provided a convenient, albeit crude, method for characterizing neighborhoods according to residents' income, social status, and ethnicity.[24] Although race was never a factor in the delineation of boundaries, preexisting patterns of racial segregation encouraged marketing consultants to associate certain ZIP codes with spatially clustered minorities. That an area was not uniformly black only made it easier for a firm to claim its reluctance to do business there was not racially motivated—or at least easier until community advocates denounced the practice as offensive and possibly illegal. In the mid-1990s, for instance, a locally owned auto rental company in Syracuse created a controversy by blacklisting the 13205 and 13207 ZIP codes. Because the company considered these zones high crime areas, an adult resident with a clean driving record and good credit rating was no more welcome than a chronic alcoholic or convicted drug dealer. Were the firm a bank, it would have been prosecuted, the local newspaper argued.[25] Outraged city officials retaliated by banning the company's valet-parking vans from the local airport.[26] Seeking an out, the firm swapped ZIP codes for a more detailed, allegedly fact-based map with "no rent" and "use caution" zones in other parts of the city and adjoining suburbs—real redlining but hardly a solution. Bad press and a lawsuit filed by the state attorney general eventually forced the firm to drop place-based discrimination altogether.[27]

While ZIP codes and finer-grained redlining are equally invidious as tools of locational-sorting, they are as different as a meat axe and a stiletto in the sense that the former can do more harm more quickly, more openly, and with far less thought or skill. Readily available and lacking the negative implications of color-coded map symbols, ZIP codes would seem the more dangerous threat. In 2008, with the housing industry reeling from the implosion of the subprime mortgage market, syndicated columnist Kenneth Harney, who writes about home buying, warned that "widespread designations of entire ZIP codes, metropolitan areas—even entire states—as 'declining markets' [could] hinder a real estate recovery and hurt minority groups and moderate income buyers disproportionately."[28]

Although federal fair-housing legislation might limit abuse by firms that loan money or sell mortgage insurance, state laws offer the most promising strategy for ending the misuse of ZIP codes in setting car insurance rates and determining eligibility. California is an exemplar, principally because of Proposition 103, approved by voters in a 1988 referendum. The state's insurance regulations now ban what the Consumer Federation of America denounces as "territorial rating [whereby] the vast majority of good drivers are forced to subsidize the rates of a few bad drivers located in the same ZIP code."[29] Rates now depend primarily on a motorist's driving record, years of experience, and annual mileage. That Californians could vote directly on Proposition 103, rather than depend on legislators intimidated by lobbyists, reflects their state's tolerance of public referenda—one of many consequences of the jurisdictional boundaries examined in chapter 6.

Greenlining

No one writes about the positive impact of HOLC maps on a city's green, A-list neighborhoods, but anyone who did might appropriately call it *greenlining*. Far less common than its evil opposite, the term aptly describes place-based approaches to economic development devised by federal and state agencies in the 1980s and touted as "enterprise" or "empowerment" zones.[30] While the latter adjective suggests a concerted effort to boost the self-confidence as well as the political and economic clout of local residents, government-sponsored greenlining ventures almost always focus on subsidizing businesses as an incentive for creating jobs. As a strategy for combating unemployment, underemployment, or out-migration, greenlining typically involves mapping out areas within which firms that create jobs receive tax breaks or outright grants. Zone boundaries are important because rewards accrue (in principle at least) only to employers who relocate or start a new business within an enterprise zone, or expand existing operations there. Local entrepreneurs benefit if the venture is successful, and nearby residents benefit when they're hired.

Place-based economic redevelopment has been neither a clear success nor an utter disaster. Reliable assessment is handicapped by a glut of programs at both state and federal levels as well as the difficulty of gauging effectiveness. More than 40 states established enterprise zones, starting in the early 1980s, but the diversity of settings, incentives, and experiences defies straightforward evaluation. Planning scholars Alan Peters and Peter Fisher, who examined 75 zones in 13 states, could only conclude that some programs worked but most didn't. "We do not have good measures of what government is giving away in enterprise zones," they grumbled, and "we don't know enough about the size of the public carrot being offered to the private sector."[31]

Federal place-based programs, initiated in 1993, are more homogeneous yet equally resistant to evaluation. In a 183-page study released in 2006, the Government Accountability Office concluded that overall the zones "showed some improvements, but our analysis did not definitively link these changes to the program."[32] Moreover, inadequate data and vague monitoring guidelines thwarted effective oversight by the federal, state, and local agencies responsible. Another shortcoming is a failure to select truly needy sites. According to public policy analyst Robert Greenbaum, "the programs appear not to be targeting areas already lacking employment opportunities."[33] And even when an enterprise zone is within or next to a disadvantaged inner city neighborhood, there's no guarantee that local residents will benefit from whatever new jobs arise. Walking distance might have been important in nineteenth-century factory towns, but it was largely irrelevant in late twentieth-century America, where even low-wage workers might have a commute of 30 miles or more. This spatial disconnect led urban economists Barclay Jones and Donald Manson to argue as early as 1982 that "the enterprise zone concept is unworkable on a large scale in the United States."[34]

Whatever their role in empowering distressed neighborhoods, empowerment zones have been a magnet for greedy companies eager to accept subsidies but unwilling or unable to create new jobs. *Greedy* might not be the right word—*dishonest* or *cynical* would fit just as well, judging from an ongoing expose in my local newspaper, which questions the management, if not the wisdom, of New York State's Empire Zone (EZ) program—in the lexicon of state nicknames, we're "The Empire State." Created in 2000 to encourage new businesses as well as dissuade companies from transferring work out of state or offshore, the EZ program grew to channel subsidies through more than 80 zones statewide.[35] As expansion progressed, benefits began to flow to firms that didn't need them, even companies not inside a zone.[36]

Tax reduction credits can be exceptionally lucrative for small employers. If a company with a single employee adds a second worker, the resulting 100 percent increase triggers a 100 percent reduction in its state income tax, which means it pays nothing for 10 years.[37] According to the *Syracuse Post-Standard*, this "loophole" helped NRG Energy, a New Jersey company that owned three "old power plants that either rarely operate or are among the state's worst polluters," claim more than $20 million in tax credits in 2006.[38] Consistently at or near the top of the "tax-breaks" list, NRG "employs fewer than 10 workers" and is a notorious "shirt changer." As another front-page story explained, firms that "shifted their assets to spin-off companies or filed new incorporation papers earned the nickname . . . because it seemed like they simply asked their employees to change their shirts from Company A to Company B and return to their jobs."[39] EZ money also went to politically connected law firms, which are unlikely to decamp for the Sun Belt, and large national chains like Lowe's and Wal-Mart, known for driving local firms out of business.

Failure had few consequences. Companies that missed their job-creation or investment targets were rarely kicked out of the program, while firms that expanded their plants and job rolls as promised were dropped because of a technical detail or paperwork snafu.[40] While state officials promised reform, James Parrott, chief economist at the Fiscal Policy Institute, a not-for-profit watchdog organization, charged that the program "has strayed so far from its original, well-intentioned purpose, it has become a source of ridicule [and] should be scrapped."[41]

Whatever their misgivings, local governments responded enthusiastically by filling out detailed applications that described plans to offer child care as well as to train zone residents for new jobs. According to the state's 73 pages of regulations, a county was generally entitled to 1,280 acres of

EZ territory, which could be divided among several "distinct and separate contiguous areas," each with a census tract or cluster of blocks above average in poverty level or unemployment.[42] Even so, new hires did not have to live within the zone, and as long as the total area did not exceed the municipality's allocation, its zone coordinator could extend EZ benefits to comparatively small parcels outside the larger "distinct and separate" areas.[43]

Critics of the program were particularly outraged when several small municipalities that received as many EZ acres as much larger cites sold their surplus EZ benefits to developers beyond their borders. Geneva, a small city in western New York, had the only Empire Zone within Ontario County. After covering most businesses within the city limits as well as a few others in an adjoining town, Geneva still had a substantial surplus, which it farmed out to a shopping mall in the Town of Victor, 30 miles away but still within the county.[44] In return, Geneva received $150,000 a year and the right to install a kiosk at the mall advertising its industrial and tourist attractions.[45] In a similar exchange of tax credits for cash, Potsdam, a village in northern New York, sold 35 EZ acres for about a million dollars a year to Reliant Energy, a Texas firm operating 21 small hydroelectric power plants in the region. When Reliant sold the plants, its EZ credits transferred to the new owner.[46]

Empire Zone regulations mention cartography only once, in requiring that applications include "any maps . . . of the area comprising the proposed empire zone, showing existing streets, highways, waterways, natural boundaries and other physical features."[47] Despite the state's nonchalance, local officials use a variety of maps to pitch their zones to prospective employers, few of whom would forego EZ benefits. Onondaga County, where I live, supplements an overview map locating its five "distinct and separate areas" with comparatively detailed individual maps for each area and a countywide map identifying EZ certified businesses outside these five subzones. Most beneficiaries are small firms. For example, the excerpt in figure 8.3 includes Quarryside Animal Hospital, which claimed $6,851 in tax benefits on its 2005 state tax return but didn't meet its job and investment goals. I know these details only because the *Post-Standard* had sued the state to get the data, which it posted on its Web site.[48] A mile and a half south in the hamlet of Jamesville are another 11 EZ beneficiaries, identified only by parcel number on a pop-up inset map.[49] As far as I can

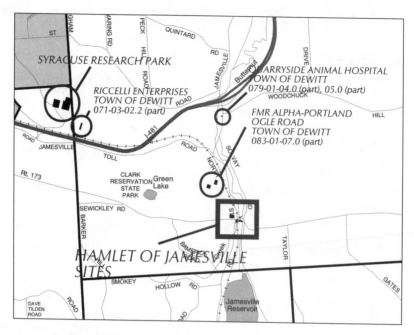

FIGURE 8.3. Excerpt from the Onondaga County Empire Zone Sites Boundary Revision, January 2003, map prepared by the Syracuse–Onondaga County Planning Agency.

tell, few are new and most are operating pretty much as they did before the program started.

....

Similarities between greenlining and redlining are as significant as their differences. Both practices use maps as a tool for distributing benefits and liabilities. Place-based discrimination favors businesses within an enterprise zone as well as homebuyers in green and blue neighborhoods, often situated in growing suburbs, where federally guaranteed loans were comparatively abundant. Less well treated, abusively perhaps, were businesses outside an enterprise zone and people eager to buy into a red or yellow neighborhood. Revenue lost through poorly managed greenlining projects must be recovered from other taxpayers, who also pay twice for redlining—once to subsidize schools and transportation infrastructure for the suburbs, and again to rehabilitate inner city neighborhoods, allowed to deteriorate because investment was diverted elsewhere. With both redlining and greenlining, maps are distractions that divert attention from the prospective borrower's ability to repay a loan and the motorist's driving record, and from a business's effectiveness in increasing investment and creating jobs. Maps, sad to say, are complicit when place-based discrimination replaces performance-based discrimination.

More intriguing are the differences. Redlining is roundly condemned, in no small way because of the inherently abhorrent rhetorical symbolism of the HOLC maps, which are widely perceived as more influential than the comparatively mundane FHA *Underwriters Manual* and Robert Park's theories of urban ecology, which underlie the rampant ethnic and racial discrimination in the housing market from the 1930s into the 1960s, and perhaps beyond. I doubt that ZIP code maps, which seem an even more powerful tool of mindless geographic discrimination, will ever be so roundly vilified. As for greenlining maps, their fluid boundaries might one day be appreciated as a cartographic enigma, useful in convincing gullible taxpayers that place-based (rather than performance-based) economic development was worthwhile. Draw a boundary on a map, stick a label on it, and people think it's real.

CHAPTER
NINE

....

GROWTH
MANAGEMENT

Few maps are as powerful as zoning maps. If you don't believe me, ask Roger Scott, a feisty trial lawyer who ignored land-use rules in Skaneateles, New York, by building a house more than twice the size allowed. In 1991, after years of litigation, the town tore down his 11,000-square-foot lakeside home and billed him for the demolition. A bizarre chain of defiant antics led to Scott's disbarment, but he was back in the news years later, after applying for work (unsuccessfully) as a neighboring town's code-enforcement officer.[1] That zoning disputes rarely have such tragicomic endings attests to the clout of maps drawn up to regulate land use—people might fight over a classification but generally they respect the lines. This chapter probes the defensive cartographies that protect neighborhoods from noise, industrial explosions, and visual blight. Issues explored include historic preservation and the fluid definition of wetlands.

Threats and Restrictions

Zoning maps are one of a quartet of maps used to control a city's destiny. Foremost is the *official street map*, which frames development with a plan for extending and widening existing streets; in forecasting the acquisition of rights-of-way the official map gives land owners both guidance and an incentive for improving their property. The other three basic maps are a *current land-use map* showing how the city looks now, a *master plan* describing how planners would like it to look several decades hence, and a *zoning map* with lines and regulations designed to make the next cycle's land-use maps resemble the current master plan.

As a regulatory tool, the zoning map reflects police powers delegated to the municipality by a state's constitution or legislature, which also gives local officials the power to enforce building codes and environmental regulations, collect property taxes, and compel (with fair compensation) the sale of land needed for public use. New York City pioneered citywide zoning in the United States in 1916, with the approval of the state legislature. George Ford, an architect and planner who had lobbied legislators and helped draw up the city's innovative zoning maps, put out an urgently persuasive plea in *The American City,* a monthly magazine for municipal administrators:

We are rapidly coming to recognize in building our cities and towns that, despite our fetish of the inviolateness of private property rights, it is poor policy to let every man develop his property just as he pleases. For so very often it is done in a way that not only is poor business for the owner himself, but highly detrimental to all his neighbors. Or, again, an utterly selfish and grasping owner will go into a settled district and put up an entirely different sort of building, that depends wholly for its light, air and attractiveness on what it steals from its surroundings. It is a parasite. Its value depends on the other buildings remaining as they are. The moment the surrounding properties are developed like his, his value is gone. Districting would keep the vampire from sucking the life blood out of his neighbors.[2]

Frightened by images of gasoline stations, fuel storage tanks, and small factories wedged next to otherwise attractive single-family homes, other cities followed New York's lead.[3] But many more municipalities were reluctant to adopt zoning until 1926, when the U.S. Supreme Court ruled that Euclid, Ohio's zoning ordinance, intended to separate incompatible land uses, was "a valid exercise of authority."[4] Zoning might limit the owner's use of a land parcel, the court noted, but a well-crafted zoning law is not an unconstitutional "taking" of the owner's property right. Citizens can challenge a zoning map's delineations, notes zoning guru Donald Elliott, an attorney and city planner, but "as long as the city's determinations have a reasonable basis related to some legitimate government interest, its line drawing will almost always be upheld."[5]

Like other pioneering legislation, New York's zoning ordinance was an anxious reaction to an unforeseen nuisance, in this case the seven-acre shadow cast over lower Manhattan by the Equitable Building, newly constructed on a full city block with no setback whatsoever and reaching upward an unprecedented 36 stories thanks to advances in elevator technology.[6] To thwart greedy developers, planners drew up a trio of maps that partitioned the city among use districts, height districts, and area districts. The use-district map, which divided the city into residential, business, and "unrestricted" districts, focused on keeping noise, noxious odors, and traffic out of purely residential neighborhoods. Because industry could not be banished altogether, unrestricted districts were set aside for manufacturing and other pariah activities. Addressing aesthetic concerns, the height and area districts promoted a sunnier, more open cityscape by capping a building's height or pushing its outer walls back from the property line.[7]

Instead of a rigid citywide cap on architect's vertical aspirations, the height-district map made maximum height a multiple of street width and divided the city among five categories. In the "1½ times" district, for instance, the front wall of a building on a street 80 feet wide could rise to 120 feet. Extra floors were permitted, but only if the exterior wall was moved a foot farther inward for each additional three feet in height—the recipe for a spate of "setback skyscrapers" with the classic multitier, wedding-cake profile of the Chrysler Building. Area districts further narrowed the "building envelope" by specifying ground-floor setbacks and limiting the portion of the site occupied by the structure. Restrictions were most severe in upscale residential neighborhoods in the city's outer boroughs, and most permissive in lower Manhattan's financial district.

An excerpt from the use-district map sheet covering Midtown Manhattan (fig. 9.1) reveals a strategy of putting opposite sides of the street in the same district, lest obnoxious activity in plain sight sully a more genteel environment. Unlike most contemporary zoning maps, which typically rely on color coding to show a more complex array of categories, New York's original use-district map was printed in black-and-white. The most prominent symbols here are the solid black lines representing business streets, especially prominent in Midtown. Next most apparent are

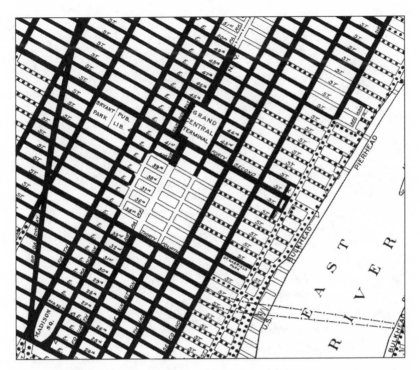

FIGURE 9.1. Excerpt for part of Midtown Manhattan, as shown on New York City's Use District Map, Section 8, July 25, 1916. Solid black, dotted, and blank (white) lines represent business, "unrestricted," and residential uses, respectively.

the dotted lines representing "unrestricted" uses, in this case industrial or mixed industrial activity extending a block or two inland from the East River, lined with piers early in the century, when marine transportation and dockside manufacturing were particularly important. In the heavily developed Midtown area, white lines representing residential streets stand out as oases of affluent tranquility in the Murray Hill neighborhood south of Grand Central Station and in two other clusters of "long blocks" along comparatively narrow streets running east to west. By contrast, Manhattan's less numerous north-south arterials, widened to accommodate trucks and streetcars, were more suitable for shops and office buildings than the generally narrower east-west side streets.

Today's Midtown zoning map is radically different. The most dramatic change is the comparatively complex taxonomy reflecting a more nuanced approach to city planning as well as the consolidation of use, height, and area restrictions into a single set of districts. Close inspection of the current

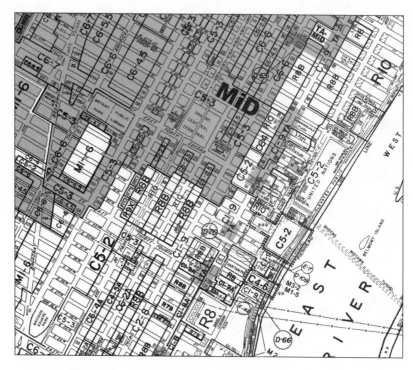

FIGURE 9.2. Excerpt from the contemporary New York City zoning map. Area shown corresponds to the geographic scope of figure 9.1 and measures approximately 1.5 miles from left to right.

zoning map (fig. 9.2) for the same section of Midtown reveals no fewer than 39 zoning districts (if one peeks beyond the excerpt's edges). Districts are identified by a letter (C, M, or R for Commercial, Manufacturing, or Residential) followed by one or more numbers or letters that link a name describing the district's intent to a catalog of specific restrictions. Among the commercial categories shown here, C1 and C2 indicate Local Retail and Local Service Districts, respectively, while C4, C5, and C6 are General Commercial, Restricted Central Commercial, and General Central Commercial Districts. An alphabetized list of specific uses indicates that department stores are inappropriate for C1 and C2 districts, while costume rental is banned from C1 areas and pawn shops from C1 and C5 areas. Not surprisingly, manufacturing is confined here to a few sites along the river and several substantial M1 ("Light Manufacturing") areas south and west of Bryant Park in the fabled yet fading Garment District.

Residential zoning in Midtown is less straightforward. Although all of the eight R districts on the map excerpt are considered General Residence Districts, they differ in ways meaningful mostly to architects, engineers, and real estate developers willing to wade through the 2,784-page *Zoning Resolution*, which I downloaded from the City Planning Department's Web site as a 29.4-megabyte PDF file. The most prominent difference involves the floor-area ratio (FAR), calculated by dividing a building's total floor area by the area of its lot. In R8B districts, which cover the interior portions of many east-west streets, the FAR may not exceed 4, while in the R10 district along the East River north of the United Nations the maximum FAR is 10. The clear intent is to protect mid-block town houses in established East Side residential neighborhoods but encourage high-rise apartment buildings surrounded by open space and parking lots on former industrial land along the river.

Parts of the city receive the added protection and encouragement of an overlay of "special purpose" districts, shown on the map by gray shading and bold letters. The excerpt includes all or portions of five of the city's 34 special districts.[8] The smallest is United Nations Special Development District ("U"), intended to promote a "unified design concept" on two tracts near the UN Building by raising the maximum FAR from 10 to 15. Equally distinctive is the Special Transit Land Use District ("TA"), created to foster openness around likely entrances to the long-awaited Second Avenue Subway, first proposed in 1929. A square patch of darker shading toward the upper right marks the coincidence of a small TA district and the much larger Special Midtown District ("MiD"), created to coordinate planning for the Midtown business district, including Fifth Avenue and the Theater District, just west of Times Square. Concerns included congested pedestrian traffic on principal shopping streets, unwanted commercialization along residential side streets near the Museum of Modern Art, enormous animated signs in Times Square, and legitimate theater's fragile foothold in a neighborhood where developers were eager to buy and demolish historic structures. The city's *Zoning Resolution* lists 17 distinct goals ranging from "improving working and living environments" to "conserv[ing] the value of land and buildings and thereby protect[ing] the City's tax revenues."[9]

A close reading of the zoning map reveals additional land-use limitations such as "restrictive declarations," agreed to by the owner and binding on all future owners. Tiny arrows connect each affected parcel to a circle inscribed with a number (like "D-109," just left of the S in "EAST RIVER") referring to a specific covenant, achieved with deeds and other property records. Similarly numbered "environmental review designations" (like the "E-7" directly above the D-109) flag restrictions related to hazardous materials, noise, or air quality.[10]

Missing from the map are "variances" granted by the city's Board of Standards and Appeals, which can give a land owner "relief" from the "practical difficulty" or "undue hardship" that could result from strict enforcement of zoning regulations.[11] Variances require formal notification of neighborhood residents and owners of nearby parcels as well as a public hearing, at which neighbors and other interested citizens can argue for or against a formal waiver. Maps and aerial photos have an important role at these hearings, to locate the property in question, explain the need for a variance, and describe plausible effects on nearby parcels. Variances are not shown on the zoning map because they don't change its lines or categories, and would add needless clutter.

Securing a minor variance for a Midtown property can be expensive and time consuming, but in DeWitt, New York, near Syracuse, the occasional request to ease side-yard setback restrictions so that a homeowner's new room can be a foot or two wider are dealt with sympathetically, expeditiously, and prudently (I think) by the Zoning Board of Appeals, whose unpaid members are appointed, not elected. By contrast, "zoning amendments," also called "rezoning" because they involve relocating district boundaries or changing categories, require that a municipality's elected city council or town board formally approve a revised zoning ordinance. In big cities or small towns redrawing the zoning map is no small matter. Although a municipal legislature could, in principle, change a single lot's zoning category, the courts tend to frown upon preferential "spot zoning."[12]

When DeWitt began updating its zoning map several years ago, town officials paid special attention to the hamlet of Jamesville. (In New York a hamlet is like a village but generally smaller and not incorporated.) Jamesville is proud of its origins as a nineteenth-century mill town that got an added boost in the 1850s, when the railroad not only reduced travel time into Syracuse but made it economical to heat homes with clean-burning anthracite from Pennsylvania. Railcars and coal took on ominous overtones in late 2006, when an energy entrepreneur proposed building a coal gasification plant on the site of a defunct cement plant north of the hamlet's business center. Each day the plant would convert 100 carloads of coal into 1.8 million therms of methane heating fuel and create as a side product six tank cars of sulfuric acid—a valuable industrial chemical but nasty stuff when spilled.[13] Although the plant would enhance the tax base and create at least 100 permanent jobs, local residents feared gas explosions and train wrecks as well as noxious fumes, noise, and an industrial eyesore. Lawn signs sprouted overnight—"STOP the Coal Plant"—and letters to the editor exuded the anxiety and animosity of a classic NIMBY (Not In My Backyard) dispute. That the site was literally in the backyard of the local elementary school spread fear beyond the immediate neighborhood.[14]

Fortunately for the hamlet and local parents, DeWitt had not yet completed the lengthy process of rezoning, initiated in 2004. It was comparatively easy to take the threat of a coal plant off the table (as well as off the map) by changing the site's classification. On figure 9.3, my black-and-white rendering of a portion of the old, color-coded zoning map, the former cement plant is marked by the dark "IND" (for Industrial) area at the top right—as apparent from the adjoining R-1 (medium-density residential) lots, the coal plant clearly would have been in several residents' backyards. The new map (fig. 9.4) labels the site "H-T," for High Technology, a category clearly intended to exclude older, noisier, and smellier mechanical and chemical technologies, which not only ship a lot of material in or out but might go boom without warning. Explosions are not uncommon near the hamlet—the industrial land at the top right on both maps is a limestone quarry that supplies much of the region's crushed stone—but routine blasting never evokes fear of a toxic plume. Although the developer abandoned plans for a coal plant in Jamesville a year before rezoning became final, DeWitt's new High Technology zone signifies a continuing commitment to the hamlet's safety.

Keeping Up Appearances

Keeping obnoxious land uses at bay was not the only reason for rezoning Jamesville. A second goal—and the principal aim until the coal plant loomed large—was preservation of the "unique character of a 19th century mill town."[15] Although the amended ordinance's exact wording seems to confuse humble Jamesville with more architecturally ossified communities like Virginia City, Nevada, or Wiscasset, Maine, neither of which bears so deeply the scars of twentieth-century transport and commercial signage,

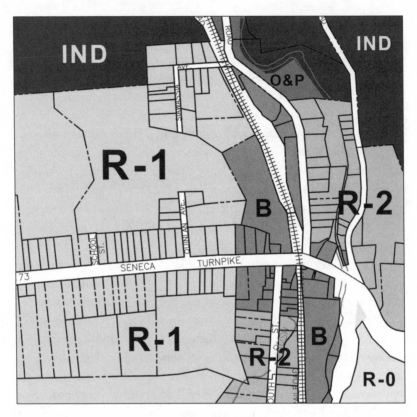

FIGURE 9.3. Jamesville, New York, on pre-2008 zoning map. Area shown measures
approximately 1.2 miles from left to right.

the intent of the Jamesville Hamlet District ("HAM" on fig. 9.4) is as understandable as its framers' insistence that "all new building[s] should incorporate architectural styles from this period" and that "site design" be
"pedestrian friendly." Anxiety over parking lots ("one tree planted for every
five parking spaces") and signage ("detached signs shall be or have the
appearance of being hand crafted from wood or stone") no doubt reflects
the Hamlet District's incorporation of the old Business District ("B" on
fig. 9.3).

Although the single-family dwelling is the preferred type of new construction, 28 other uses are permitted, subject to a "site plan review." The
"permitted" list includes bed-and-breakfasts, offices, and animal hospitals,
but bans "drive thru" banks and restaurants. Sorry, McDonald's. Uses allowed in a Business District but banned from the Hamlet District include

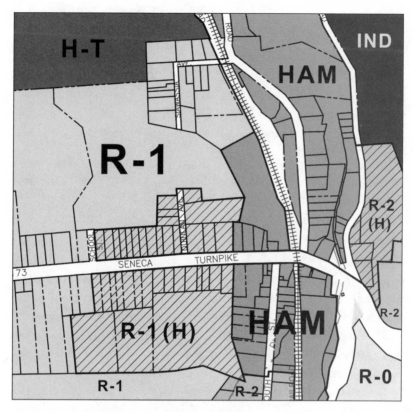

FIGURE 9.4. Jamesville, New York, on new zoning map. Area shown corresponds to geographic scope of figure 9.3 and measures approximately 1.2 miles from left to right.

animal day care, commercial garages, vehicle sales, and telecommunications towers. To avoid undue hardship to current business owners, the new rules don't affect existing operations as long as the structure and its use remain unchanged—like most municipalities, DeWitt prefers "grandfathering" to ill will and costly litigation. In short, the local body-and-fender repair shop can remain as long as it cleans up its act and doesn't try to expand. If DeWitt's strategy of forward-looking zoning and slow attrition works, significant improvements in the hamlet's appearance will attract investors willing to buy out and replace nonconforming businesses. To paraphrase the mantra from *Field of Dreams,* if you zone it, they will come.

Visual design is no less important than land-use restrictions. Because the Hamlet District contains vacant land as well as a few structures ripe for replacement, design standards promoting a bygone (if not mythical)

FIGURE 9.5. New primary buildings in the Jamesville Hamlet District must have three of the five architectural features shown here. "All or any combination of" cornice brackets, fanlight windows, window shutters, and bay windows count as a single feature.

landscape insist that "the dominant exterior finish shall be *or have the appearance of* horizontal wood siding or natural stone" (emphasis mine). Equally prescriptive are the requirements that "buildings shall be [*sic*] or have the appearance of a pitched roof" and include "at least three" of the five preferred "architectural features" juxtaposed visually in a diagram that exudes monstrous quaintness if studied for more than half a minute (fig. 9.5). Insofar as all historic districts depend on a measure of aesthetic conformity, Jamesville's cafeteria of clichés is hardly unreasonable.

A second component of the planning board's vision for Jamesville is the Hamlet Residential Overlay District, delineated by the superposition of closely spaced parallel lines on parts of existing R-1 and R-2 districts, and identified on my black-and-white map (fig. 9.4) by the addition of "(H)" for Hamlet Overlay. Following a standardized zoning schema used throughout the country, DeWitt's R-1 and R-2 districts are residential zones intended principally for single-family homes and unobtrusive in-home businesses like law offices and small day-care centers—the key difference is density, with slightly larger and wider lots in R-1 areas. To preserve the neighborhood's residential character, the overlay imposes additional conditions on existing districts otherwise left intact. As in the Hamlet District, nonconforming uses or structures may not be expanded. In addition, alterations must "be consistent with the design of the original structure," and new and rebuilt structures must conform to design guidelines similar to the Hamlet District requirements. Curious differences include the addition of brick as an acceptable exterior finish and the removal of cupolas from the menu of mandated architectural features (fig. 9.5).

DeWitt's plan for Jamesville is as much about redesign as it is about historic preservation. Because rezoning is a political process, town officials are acutely aware that changing how a community looks and functions is potentially divisive. A more proactive approach, with aggressive weeding out of nonconforming uses and tighter controls on building materials and house color, might promise quicker results, but success depends upon recognizing that any new map must be sold to residents rather than imposed by professionals. For these reasons, the planning board involved the community in the process from the outset through public hearings and a local Hamlet Zoning Committee with a broadly inclusive membership. And when committee members presented the new map at a final information hearing, they emphasized that the plan "address[ed] the future of the Jamesville area while having zero effect on current residents and businesses."[16] The only change following an otherwise uneventful presentation was an amendment waiving the design standards for an owner eager to repair or replace a grandfathered dwelling severely damaged by fire— insurance settlements don't cover Victorian makeovers.[17]

With minimal impact on current residents, Jamesville's hamlet zoning is Historic Preservation Lite, in contrast to the industrial-strength variety found in Washington, DC's Georgetown or Philadelphia's Society Hill. It's doubtful that the Jamesville Hamlet District, now or years hence, could qualify for the National Register of Historic Places. As defined by the National Park Service, which administers the Register, a *historic district* is "a geographically definable area, urban or rural, possessing a significant concentration, linkage, or continuity of sites, landscapes, structures, or objects, united by past events or aesthetically by plan or physical developments."[18] I've seen numerous federally recognized historic districts, and Jamesville is (to be kind) noticeably more eclectic. And even if it were listed, the Register would afford little protection, unless perhaps federal highway planners threatened to insert an interstate highway. Because federal registration wards off only federally financed incursions, heavy-duty historic preservation is almost always endorsed by the local legislature, which holds the real power to regulate land use. Communities with a stronger sense of architectural or historic heritage than DeWitt can delegate the rigorous administration of an exhaustive litany of architectural restrictions to a local committee of zealous preservationists who vigilantly micromanage homeowners eager to repaint their houses, resurface their driveways, or reshingle their roofs.[19] While aesthetic results might justify rigorous enforcement, it's easy to understand local resentment.

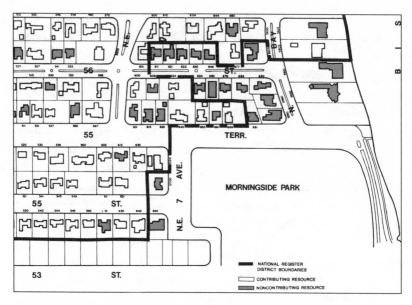

FIGURE 9.6. Portion of the boundary of the Bay Shore Historic District, in Miami, Florida.

Resistance to historic district zoning stems largely from contiguous districts that are neither well defined nor uniformly significant.[20] State historic commissions, many of which have their own registries, recognize federal guidelines that distinguish between "contributing resources," with sufficient significance to warrant protection, and "non-contributing resources" that lack architectural gravitas but enhance the district's compactness.[21] As illustrated by the boundary for Miami's Bay Shore Historic District (fig. 9.6), lassoing all the contributing resources often means incorporating noncontributing neighbors, some of whom would rather retain the right to install vinyl siding or paint their houses purple. Some municipalities let a majority of local property owners veto historic preservation, but most don't.[22] Boundary delineation can become particularly contentious if preservationists, sensing the threat of deterioration, widen the visual buffer around a core of significant structures. Members of the historic district committee or "landmarks commission"—already conspicuously elite—must work at not appearing arrogant. Should they fail, resentful residents understandably dig in their heels.

Not all owners resist or complain.[23] Some are mollified when tax incentives fund repairs and reverse visual blight. Others enjoy the prestige of living in a historic district, which can enhance property values, even for noncontributing structures.[24] Indeed, property owners left outside the boundary sometimes protest their exclusion. Owners, of course, are not the only residents affected. In declining neighborhoods, historic preservation is often a precursor to gentrification, with devastating consequences for nonowners forced by rising rents to move elsewhere. Focused on private property and land owners, the zoning map can be a powerful instrument for displacing or excluding the urban poor.[25]

Wet Edges

Land-use controls also focus on water, which is hardly surprising given current concerns about clean drinking water, flooding, and biodiversity. The complexities of water use and the hydrologic cycle require a variety of map-based regulations. Flood maps help floodplain managers control development in "regulatory floodways" adjacent to the channel, where structures are not only readily damaged but likely to interfere with the expeditious discharge of floodwaters.[26] Geologic and soils maps point out areas where a high water table, highly permeable soil, or exposed bedrock makes groundwater especially vulnerable to underground fuel tanks, land-fills, and septic tanks.[27] Zoning maps lend a hand by requiring larger build-ing lots in subdivisions served only by septic systems, banning hazardous land uses in industrial and business zones, or limiting impervious surfaces like paved roads, rooftops, and parking lots, which shed rainwater rapidly and interfere with normal recharge of groundwater. Planners are especially wary of expedited storm drainage, which can be slowed by marshes and other wetlands, also valuable as wildlife habitat. Development is a threat to wetlands, which can be drained or filled in, but protective legislation calls for specialized maps that override whatever permissiveness a zoning map might imply.

Wetlands are among the most difficult topographic features to map. Although *wetland* emerged early in the twentieth century as a euphe-mism for swamp, the term acquired a more nuanced meaning in the 1970s, when scientists and public officials recognized the ecological importance of areas with permanently or periodically saturated soil or a water table exceptionally close to the surface.[28] Tidal wetlands, though vulnerable to severe storms, coastal erosion, and sea level rise, are generally more stable and easier to delineate than freshwater wetlands, which seem to vary in size from day to day, season to season, and year to year. Is that shallow depression in the backyard a wetland because it floods for an hour or so after an exceptionally heavy rain? Probably not, but before bringing in a backhoe or applying for a building permit, it's wise to call in a professional wetland delineator to check the soil, vegetation, and terrain for "wetland indicators" recognized by regulatory agencies.[29] As wetland ecologist Ralph Tiner notes, "a wetland is whatever the law or government regulation or local zoning ordinance says it is."[30] A certified delineator, who typically has a degree in soil science or plant ecology, will check soil moisture at various depths as well as look for flora on the National List of Vascular Plant Species That Occur in Wetlands.[31] Because wetlands enjoy the protection of state and local laws as well as the Clean Water Act, enforced by the U.S. Army Corps of Engineers, determining whether a permit might be needed is a serious matter.[32] Fines for destroying wetlands typically run to thousands of dollars. In addition, regulators might order the restoration of the affected site or the creation of a substitute wetland similar in size and character.

Seasonal drought and climate change are not the only factors undermining a stable wetland-upland boundary. Federal guidelines differ from state criteria, which vary from state to state, and local regulators can adopt the federal or the state definition or compose their own. More troubling, federal guidelines have changed several times. Legal scholar James DeLong, a persistent critic of federal environmental regulations, labeled the Corps of Engineers definition a "moving target" after counting six fundamental revisions between 1986 and 1994.[33] And further change seems likely. In 2006, after a Michigan landowner challenged his conviction for filling in federally protected wetlands, the Supreme Court suggested, in a 5–4 decision that sent the case back to a lower court, that the Clean Water Act might cover only wetlands physically linked to a navigable waterway.[34] Although environmentalists think it sufficient that a wetland is merely part of a hydrologic system that includes the legally momentous "waters of the United States," lobbying by property developers, cattle ranchers, and coal producers persuaded regulators to issue new rules, in 2007, that reduced the scope of federal wetlands protection, at least for remainder of the Bush administration.[35] Stay tuned.

Lobbying is one way to change the rules, fudging delineations is another. In perhaps the most odd-ball federal wetlands prosecution on record, a

Georgia wetlands delineator named, ah, Todd Ball received a $90,000 fine and three years probation for falsifying a North Carolina wetlands map and forging the signature of a Corps of Engineers official.[36] His crimes included violating the Clean Water Act, making false statements, and conspiracy, presumably with the unnamed developer who built a golf-course access road across a protected wetland. While the current owner was required to mitigate the damage, and Ball had to pay $20,000 additional as restitution—and agree never again to work on wetland maps—the conspiring developer was not prosecuted. Go figure.

CHAPTER
TEN

....

VICE
SQUAD

Make a list: what wouldn't you want to live near? Unprompted, most of us would list busy highways and chemical plants. Yet prompted by a checklist, a clear majority of homeowners would surely add strip clubs and adult bookstores. Attitudes toward the private behavior of consenting adults might have changed considerably since the sexual revolution of the 1960s, but windowless buildings with tacky signs are an undisputed threat to property values and neighborhood ambiance. This seemingly puritanical desire to distance ourselves from sexually-oriented businesses could be dismissed as class bias, but renaming the Bada Bing a "Gentlemen's Club" and adding gourmet food and valet parking would hardly win the applause of would-be neighbors at a public hearing—even those eager to treat business clients to upscale adult entertainment. Not in my backyard, they'd scream. In addition to exploring the map's role in restricting sex businesses, this chapter examines cartographic controls on convicted sex offenders, high on any parent's NIMBY list.

Without the First Amendment, nude-dancing clubs and adult book-stores would be as easy to veto as drive-thru restaurants and scrap yards. Although the free artistic expression flaunted onstage or in a centerfold does not enjoy the same protection as political speech, municipal officials are leery that a judge might declare their zoning ordinance too vague, overly broad, or otherwise unconstitutional. Zoning officials need not be reminded of *Schad v. the Borough of Mount Ephraim*, the 1981 Supreme Court decision that struck down a small New Jersey community's total ban on all forms of live entertainment, enacted after a local bookstore installed coin-operated booths from which the customer could watch a live nude dancer through a glass panel.[1] Borough officials had hoped (they claimed) merely to reduce parking, policing, and refuse problems by focusing the local business district on the area's "immediate needs," but the High Court doubted that nude dancing or other kinds of live entertainment would create new problems.[2] Although Mount Ephraim's law was unwarranted as well as overly broad, two of the justices suggested, in a concurring opinion, that a purely residential community might legally prohibit all commercial activity, sex-related or otherwise.[3] Broad restrictions can be acceptable when there's a good reason.

Zoning is a common solution. Some municipalities confine adult uses to a specific part of town, geographically akin to the red-light district of earlier times. Others prefer to disperse adult establishments, typically keeping them at least 1,000 feet apart lest a concentration attract prostitutes or drug dealers as well as jeopardize the city's image and property values. Distancing restrictions also keep adult uses away from churches, schools, and similarly sensitive locations. American Planning Association (APA) researchers who reviewed detailed studies for over a dozen cities found that the negative effects of sex-related business, which can be highly significant within 500 feet, are substantially reduced by a 1,000-foot separation, and even by a 1,500-foot separation along the same street.[4] "Beyond 1,000 feet, there may be some impact," they concluded, "but beyond 1,500 feet there is no basis for believing there will be any impact on property values."[5]

Studies like these are important to city officials, who rely on the well-documented experiences of other municipalities to demonstrate that their zoning code is neither arbitrary nor unjustified. Distance restrictions that limit a strip-club operator's choice of locations might have the secondary effect of hampering the dancers' artistic expression, but they don't trump a municipality's legitimate interest in protecting property values and controlling crime. As long as distancing restrictions and zoning categories give sex businesses what legal scholars call an "alternative avenue of communication," they should survive a constitutional challenge.[6]

Zoning maps provide a ready-made strategy for banishing sex businesses to less sensitive parts of town. In DeWitt, New York, our new zoning law not only confines adult uses to the town's industrial districts but also demands a minimum 1,000-foot separation from incompatible uses spelled out in a nearly exhaustive laundry list:

1) A lot on which there is another adult use.
2) Any Residential District, Office and Professional District, Business Transitional District, Special Business Transitional District, Business District or High Tech District.
3) Any property that is used, in whole or in part, for residential purposes.
4) Any church or other regular place of worship, community center, funeral home, library, school, nursery school, day-care center, hospital or public park, playground, recreational area or field.
5) Any public buildings.
6) Any hotels or motels.

Because judges frown on vagueness, no lot can contain more than one adult use, and all distances must "be measured from lot line to lot line."[7] In addition, adult establishments must be licensed as well as designed so that, visually at least, what happens inside stays inside.

Licensing is crucial to effective zoning, which requires continuing surveillance. Some adult businesses conscientiously comply with local codes while others push the limits on site and in court. Live-entertainment operations that encourage physical contact between dancers and customers are especially problematic. According to Eric Kelly and Connie Cooper, who prepared the APA report, "touching businesses, including lap dancing and nontherapeutic massage, are not protected by the First Amendment [and] businesses that create the risk of touching can be regulated."[8] Because some customers can "clearly reach climax" during a lap dance, Kelly and Cooper contend that these performances "are much closer to prostitution (sexual services for a price) than [stage] dancing, and . . . ought to be treated accordingly under local law and policy."[9] A municipality that bans the practice can pull the permit of any club that refuses to put its dancers up on a stage, away from the audience.

Efficient enforcement and constitutional concerns demand precise definitions—13 pages of them in DeWitt's case. A glossary that addresses an exhaustive range of land-use issues includes finely tuned descriptions for Adult Arcade, Adult Use, Adult Bookstore or Video Store, Adult Cabaret, Adult Live Entertainment, Adult Motel, Adult Motion-Picture Theater, and Adult Novelty Store. To avoid the breadth and vagueness that undermined Mount Ephraim's ordinance, DeWitt defines Adult Live Entertainment as "a business where an adult male or female exposes parts of his or her body identified in the definition of 'specified anatomical activities.'" And what's *that* exactly? Farther down the alphabetical glossary are the frankly explicit details:

SPECIFIED ANATOMICAL ACTIVITIES—The display of less than completely and opaquely covered human genitals, pubic region, pubic hair or buttocks or female breast or breasts below a point immediately above the top of the areola. Human male genitals in a discernibly turgid state, even if completely and opaquely covered.

I'll spare you the next entry, for "Specified Sexual Activities." Like other municipalities, DeWitt depends on legal consultants to work out appropriate language—court-tested whenever possible.

While DeWitt chose to disperse adult uses—a strategy encouraged perhaps by an abundance of undeveloped industrial land—Cicero, the next town to the north, prefers to concentrate sleazy commerce in a rural area near its extreme southeast corner.[10] Corner siting is a convenient ploy when local officials hope to minimize controversy over a new landfill, sewage plant, or other obnoxious facility—likely neighbors might object, but most of them vote somewhere else. In Cicero's case, the town's 1,945-foot-long adult-entertainment district is along Oxbow Road, a dead-end street conveniently inaccessible from other parts of the town (fig. 10.1, *left*). To get there you must enter from the south, along Ferstler Road, through the Town of Manlius (fig. 10.1, *right*). Manlius officials, who first learned about the district in the local newspaper, were not amused.[11]

A year and a half after passage, Cicero's sex-business district had not been added to the zoning map available for inspection at town offices—

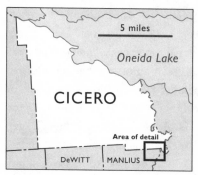
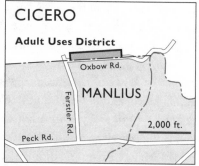

FIGURE 10.1. Cicero, New York, established an Adult Uses district in the southeast corner of the town (*left*) along a dead-end street accessible only through the Town of Manlius, just across the street (*right*).

because the law's wording alone is crucial, graphic depiction isn't required. A printed version of the recently amended zoning ordinance lays down Cicero's "Adult Uses" regulations, which account for 12 of the booklet's 51 pages.[12] Although sex businesses may not locate within 1,000 feet of a church or school, or within 500 feet of a residence, they can apparently locate side-by-side or even share the same lot. A short section headed "Location" confines adult uses to four adjacent lots listed by tax-parcel number.[13] The local tax map (fig. 10.2) thus describes the district's precise location, directly across the road from similarly vacant land in Manlius. The long, thin lots extend northward from Oxbow Road, but only the first 250 feet is authorized for adult uses. By my reckoning, the entire Adult Uses district is barely bigger than 11 acres.

I doubt that Cicero's Adult Uses statute would survive a constitutional challenge. Courts are wary of the "effective preclusion" that occurs when a zoning ordinance either banishes sex businesses to a remote, poorly accessible part of the jurisdiction or includes distancing restrictions that offer few, if any, geographic options.[14] Cicero currently has no sex businesses, and town officials apparently hope eager entrepreneurs will locate elsewhere in the region, perhaps in DeWitt, where they can more readily negotiate the necessary approvals.[15] But even if the law is constitutionally shaky, some voters no doubt applaud the effort to forestall a potential problem.

Unlike Cicero, most municipalities that regulate adult businesses prefer dispersion to concentration.[16] Perhaps the most (in)famous concentration is Boston's "Combat Zone," set up in 1974 to halt the encroachment of peep

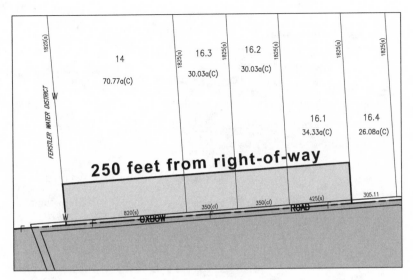

FIGURE 10.2. Cicero's Adult Uses district (light-gray area) occupies the front 250 feet of four lots (14, 16.3, 16.2, and 16,1) along Oxbow Road. The district boundary was added to this excerpt from the county tax map. Area shown measures one-half mile left to right.

shows, XXX-rated movie houses, and pornographic bookstores on Beacon Hill and other genteel neighborhoods.[17] Crime and violence quickly invaded the Boston Adult District (BAD, naturally), and city officials realized that concentration was a mistake. Zoning codes and stepped-up policing couldn't quite tame the beast, but urban redevelopment and computer technology launched an effective counterassault by raising rents and offering Internet porn as a cheaper, safer alternative. As a recent tourist guidebook noted, "The heyday of the Combat Zone has passed (although there are still a few strip clubs) [and] many of the old theatres ... were revamped for Broadway shows and upscale nightclubs."[18] Even so, the district persists on Boston's zoning maps (fig. 10.3), where it straddles the border between Chinatown (*lower right*) and the Midtown Cultural District (*upper left*).

Because the voyeur venues that once characterized the Combat Zone and New York's Times Square have largely fled to cyberspace and cable television, more than one city attorney has wondered whether the Internet might provide a constitutional substitute for bricks-and-mortar porn. Charlotte, North Carolina, raised this point in countering claims that distance restrictions in its Adult Zone Ordinance impaired free speech.

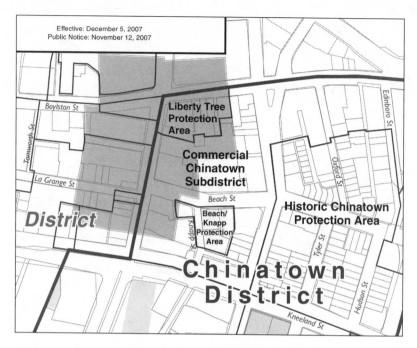

FIGURE 10.3. The gray area near the top of this excerpt from Boston's online zoning map is the Combat Zone, symbolized in light orange on the full-color Web map and identified in the key as the "Adult Entertainment District." A text block, partly visible at the upper left, indicates that the map was current in late 2007. Area shown measures one-quarter mile left to right.

The federal district judge who heard the case acknowledged that "the Internet has changed the legal and cultural landscape in obvious ways" but questioned "whether the resources available on the Internet are identical to those provided by the Plaintiffs."[19] He also doubted that "access to the Internet is as universal as access to physical structures." Even so, Charlotte got to keep its restrictions principally because its law was, as zoning lawyers like to say, narrowly tailored to serve a substantial government interest.

Buffer Zones for Pedophiles

A more menacing threat, many parents contend, is the convicted pedophile whose obsessive lust for young children is not stifled by threats of jail. Recidivism is frighteningly high, at least for level 3 sex offenders, categorized as dangerous because of a history of violence and a high risk of re-offending, and if constitutional "due process" precludes permanent confinement, continual surveillance seems a reasonable, legally defensible substitute. All states now have a Megan's Law registry, named for a seven-year-old New Jersey girl raped and brutally killed in 1994 by a neighbor who had been released several years earlier, after six years in prison for sexually assaulting another young girl. Had Megan Kanka's parents known about the pervert living across the street, she'd still be alive, legislators figured. Nowadays, no convicted sex offender can be set free, either on parole or after having served out his (or rarely, her) sentence, until a panel of experts assesses his likely danger to the community and assigns a risk category that determines the level of surveillance, which includes reporting any change of address to state officials. Local police thus know where they live and so do we, at least for higher risk registrants. In most jurisdictions, a Megan's Law Web site provides ready access to a mug shot in addition to the registrant's name, address, risk level, physical description, and criminal history (sometimes including the victim's age and gender).

Public access varies from state to state. In New York, for instance, level 1 (low risk) offenders must report their address for 20 years, while those in levels 2 (moderate risk) and 3 (high risk) are on the leash for life. An online directory lets anyone with a computer search for level 2 and 3 offenders by last name, ZIP code, or county.[20] A hyperlink takes the viewer unfamiliar with a street's location or address ranges to a zoomable MapQuest street map. By contrast, Massachusetts provides counts of level 2 and level 3 offenders by town, but lets Internet users search by last name, town, county, or postal code only for level 3 offenders.[21] There are no links to an address-mapping Web site, and viewers seeking information on level 2 offenders are advised to consult the local police department.

Not all states use three risk levels. Georgia's online Sex Offender Registry distinguishes mere "sexual offenders" from the more violent, mentally disturbed "dangerous sexual predator."[22] Curious users can download a table with 24 columns of information for over 15,000 registrants. One column provides a brief description of registrants' offenses, which run from the puzzling ("gross sexual imposition" and "crime against nature") to the unambiguously horrible ("aggravated child molestation" and rape), and other columns point out those who have "absconded" or are currently "incarcerated." (When I visited, 21 of the 79 predators were behind bars.) In addition, the Web site provides a map showing counts of offenders

FIGURE 10.4. Customized Web map of Savannah provided by
GeorgiaSexOffenders.com.

by county and lets users search for offenders, predators, or absconders by
name, city, county, or ZIP code.

Georgia's downloadable spreadsheet is fodder for GeorgiaSexOffenders.
com, a private initiative offering interactive Google Maps tailored to individual cities. As illustrated by the zoomed-in Savannah map in figure 10.4,
clickable pushpins—red for predators, yellow for other offenders—can
summon a registrant's name, complete address, photo, crime, and predator
status as well as link to additional data and a larger picture. Anxious users
can ask to be alerted whenever a new offender or predator moves into the
area. Although the official Web site maintained by the Georgia Bureau of
Investigation does not directly link to its unofficial counterpart, the latter is easily discovered by typing "Georgia sex offender" into the Google
search engine, which awards it second place in the results list, immediate
under the official state Web site. In return for providing the cartography,
Google sells advertising along the left edge of the map.

Among the cleverest sex-offender searches is the half-mile-radius option offered by Ohio's attorney general.[23] As the example in figure 10.5
shows, 19 sex offenders live at 10 locations within a half mile of 331 S. Mar-

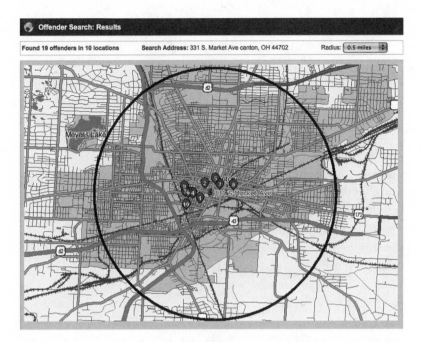

FIGURE 10.5. Web map maintained by the Ohio Attorney General's Office shows sex offenders within a half-mile of the historic Saxton McKinley House, in Canton, Ohio.

ket Avenue, Canton, Ohio—President William McKinley's former home. (To make a map I had to enter a valid street address, and the Saxton McKinley House seemed as good as any.) Each clickable symbol summoned a list of registered offenders at the numbered location, together with their names and exact address, the details of their offense, and their position on Ohio's three-level risk scale. A click away a customized map shows nearby schools, which are generally farther from the offender's residence than the McKinley House, near which they seem to be clustered. This pattern is neither an accident nor a perverse affinity for one of the country's most mediocre presidents: in 2003 Ohio passed a law prohibiting registered sex offenders from living within 1,000 feet of a school or child-care facility.[24] The distribution of schools and day-care locations throughout the city greatly narrows an outcast's choices.

Ohio is not alone. According to the Council of State Governments, between 1996 and 2007 twenty-eight other states enacted residency restrictions, typically around playgrounds, parks, and day-care centers as well as schools.[25] Distances generally range from 500 feet to 2,000 feet, but a handful of states leave the decision to a judge, parole board, or other state agency. When a new law tells a sex offender living harmlessly with relatives that he must move or be sent back to prison, family members also suffer.[26] Attorneys have protested the impact of residential restrictions on an innocent spouse or aging parents, but the courts have been largely unresponsive, despite an appalling lack of evidence that distance restrictions lessen the likelihood of repeat offenses.

Local governments sometimes impose their own bans. In New York, which (at this writing) has no statewide law, a few municipalities have established "child safety zones," configured to include churches, convenience stores, libraries, and recreation centers.[27] Around the time that Cicero revised its zoning law to relegate adult uses to its southeast corner, town officials banned level 2 and level 3 offenders from living within a mile of a school or day-care center or within 1,500 feet of the entrance to park or playground.[28] When worried residents of Salina, immediately southwest, pondered an influx of displaced pedophiles, Cicero's supervisor was unsympathetic. In a classic NIMBY response, he vented to a news reporter, "If Salina wants a law, fine. If they don't, that's fine too."[29] Left unsaid is Cicero's inability to do anything about registered offenders residing just beyond the town boundary—an oversight that New York legislators might try to address with statewide restrictions.

In addition to making sex offenders legible to Internet users, geospatial technology promises trouble-free management of residence restrictions—assuming registrants report new addresses promptly and accurately. Address-matching software makes it easy to determine a property's coordinates, used in calculating straight-line, as-the-crow-flies distance, the standard legal measurement in most statutes. For enforcement officers who prefer proactive, don't-even-think-of-living-here maps, a GIS can draw precisely proportioned buffer zones around point locations like a small day-care center and area locations like a high school or public park. Overlaying multiple layers of no-go buffers yields a map showing where registered sex offenders can legally relocate. But as Arlington, Texas's buffer map (fig. 10.6) illustrates, an arbitrarily rounded 1,000-foot restriction severely limits a registrant's opportunities when schools, play areas, and daycare centers are abundant and broadly dispersed. But because the gray areas look safer than the white areas, I'd be surprised if anxious parents just outside the safety zone aren't lobbying for a wider buffer.

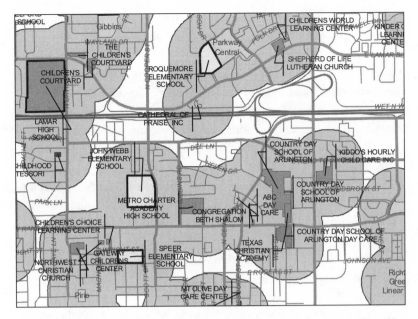

FIGURE 10.6. Thousand-foot restriction around schools, daycare centers, and parks rules out most areas in this excerpt from Arlington, Texas's North District Sex Offender Buffer Map, developed with GIS technology. Area shown measures 2.4 miles left to right.

Buffers have an ominous downside. According to a Minnesota Department of Corrections study, a proposed 1,500-foot buffer around schools "would exclude every residential area of Minneapolis and St. Paul with minor exceptions [and] force level three offenders to move to more rural areas."[30] And those who don't like country living might well join the homeless or stop registering. Fearing the unintended effects of one-size-fits-all proximity restrictions, the Department begged the legislature to continue the current policy of custom-tailored, offender-by-offender restrictions.

Ill-founded fears have led some jurisdictions to control not only where a convicted sex offender may live but also where he can work, shop, or (more pejoratively) loiter. While there is a clear rationale for banning higher risk offenders from public swimming pools and video arcades, blanket distance restrictions invite civil-liberties lawsuits, require a litany of exceptions as well as a formal process for obtaining waivers, and impose a hefty burden on police departments too poorly staffed for even token enforcement. Otsego County, New York, which prohibits level 2 and level 3 offenders from "entering and remaining" within a 1,000-foot buffer around schools, playgrounds, and recreation centers, grants waivers only after an investigation no less time consuming than the screening for a low-level government security clearance.[31] Danbury, Connecticut, which posted signs at child safety zones to advertise its ban on "loitering" by registered sex offenders, discovered a "zone hopping" loophole in its requirement that police issue only a warning the first time an offender is found inside a zone.[32] A second violation would trigger a $100 fine, but until the law was amended to read "in the same or any other" zone, a loitering pedophile could hop from zone to zone without penalty as long as he was never caught in the same zone twice.

Satellite navigation, a boon to delivery drivers and vacationing motorists, is a promising way to monitor the potentially dangerous releasee. The strategy seems simple, at least in principle: hook the sex offender to a GPS unit with an ankle bracelet and report his coordinates over the wireless network to a monitoring center for instantaneous display on an electronic map.[33] Simple perhaps, but hardly straightforward: batteries must be recharged, GPS doesn't work inside shopping malls or sports arenas, and watching a bank of computer monitors can be mind-numbingly tedious. Fortunately for employee retention and public safety, a geographic information system can automatically relate the offender's current location to no-go zones delineated on a custom-tailored map and alert officials if he enters a prohibited space, such as the home of a past victim or the minor living next door. More benignly, the system could send a wireless warning to the subject who's getting too near a child safety zone or allow time-of-day restrictions for an offender who needs to pass a playground on the way to work or the grocery store. Because detailed location histories can be archived, police can scan the registrants' movements for evidence of stalking behavior or rule him out as a suspect in a crime that occurred miles away. With Big Brother watching a releasee's every move, GPS might be an effective deterrent to new offenses. And because the offender can be gainfully employed, he can contribute to the cost of his own surveillance.

Does public safety warrant this loss of privacy for convicted offenders? How much they sacrifice and whether the public really benefits are key questions. Carefully tailored satellite tracking, which has proven effective in monitoring spouse abusers, could provide a suitably vigilant and even compassionate alternative to permanent or needlessly prolonged incarceration.[34] It's also an appropriate substitute for rigid, overly broad residential buffers, which do little good, aside from helping wily politicians persuade a gullible public that they're grappling effectively with a growing menace. Most pedophiles, sad to say, prey on relatives and household members, and because their crimes too often go unreported, they're never required to register and never told to keep their distance. Voters should be wary of buffer maps that posit deceptive solutions with negative consequences.

CHAPTER ELEVEN

....

NO DIG,
NO FLY,
NO GO

Restrictive maps often target narrow audiences uniquely affected by regulations tailored to a specific activity. Long-distance truckers, for instance, must plan routes that conform to weight and clearance limits, while hunters are told when, where, and with what kind of weapon they can "harvest" deer of a particular age and sex. This chapter explores the use of narrowly targeted maps to promote safety or conservation in diverse situations not addressed earlier. Some of these maps are straightforward, some demand specialized knowledge, and some require interpretation by specialists who mark restrictions on the landscape for ultimate users who never see, or need to see, the maps.

Call before You Dig

Because few homeowners own or rent backhoes, the typical user of these unseen maps is a contractor or landscaper. The maps, mostly electronic these days, describe buried infrastructure such as pipes, conduit, or cable carrying natural gas, water, sewage, steam, electricity, or electronic communications.[1] Excavators don't need to see the maps because the regional "one-call center," which integrates information from various utilities, can tell at a glance whether a proposed excavation is at all likely to puncture or sever an underground facility.[2] In many areas, the excavator can reach the center by dialing a three-digit number (811) heavily promoted in flyers distributed through local planning offices, equipment rental outlets, and farm supply stores. The flyers are intended mainly for do-it-yourselfers, who are a significant threat because a poorly planned excavation can trigger a disastrous fire or leave thousands of homes and businesses without telephone or Internet service for days.

Field marking makes the maps legible to the excavator. Once the one-call operator determines the kinds of facilities potentially at risk, each company or municipal department with lines near the proposed excavation sends out a crew to mark the surface above its facilities with colored flags or fluorescent spray paint. A utility that doesn't respond within two or three business days faces a substantial fine. Standardized colors include red for electric power lines, orange for communications, yellow for gas or steam, green for sewers, and blue for drinking water. (As a guide, the excavator should have premarked the approximate alignment or boundary of the proposed excavation in white.) Each utility marks its own lines, and the excavator must protect these markings as well as respect a tolerance zone at least three feet wide. Because digging is banned within 18 inches of a marked facility—except with nonpowered hand tools—width is also shown for pipes or conduit wider than two inches.

Because small errors can have huge consequences, some utility managers are rightly skeptical of their company's maps and their employees' map-reading ability. Particularly problematic is the map showing an underground facility as planned, rather than as built—pipeline and cable-laying crews that improvise or cut corners don't always record these changes on company maps. To remove uncertainty, promote safety, and lessen the likelihood of lawsuits, locator crews use ground-penetrating radar and other sensors to detect and fix the location of underground lines.[3] Electronic locators can easily detect metallic pipe, and newer lines installed with corrosion-resistant plastic pipe typically have an easily detected metal wire running along the top. While electronic detection will always be important, facilities maps have become more accurate now that GPS is used to record the location of new or replacement lines.

Getting utilities to cooperate was a significant challenge, but no more daunting than the need to register maps at diverse scales to a common grid. One-call centers caught on in the 1970s and 1980s, when geospatial technology was able to lower cost, improve accuracy, and reduce search time. Equally important were state laws requiring prior notification by contractors, homeowners, farmers, and other excavators, including the utilities themselves. Although technical standards, approved practices, penalties for noncompliance, and public-education efforts vary from state to state, arm's length map reading at one-call centers is essential to economic development and public safety.

Restricted Airspace

Few maps are as complex and rigidly regulated as aeronautical charts. Pilots need them to plan routes and locate navigation aids, and the Federal Aviation Administration (FAA) uses them to assign responsibility for air-traffic control and keep aircraft away from predictable dangers. Separate charts address the diverse needs of light aircraft, high-altitude passenger jets, and helicopters. Licensed pilots know how to read the relevant charts, and aeronautical publishers are conscientious in keeping their charts current and reminding pilots to check the government's *Notices to Airmen* for updates. Caveats abound, some a bit extreme: three-month-old surplus charts donated to university research libraries are stamped "Obsolete" in bold red letters, and sample excerpts in the *Smithsonian Atlas of World Aviation* are gratuitously labeled "not to be used for navigational purposes."[4] Cautiously apprehensive liability lawyers at work, no doubt.

In showing pilots where they may not fly, American aeronautical charts distinguish between prohibited and restricted areas, often intended to foster national security.[5] Flying is forbidden at all times within a prohibited area, except by permission of its designated "using agency." Each area is identified by a number preceded by the letter *P*, and its location is described in the *Federal Register* by the latitude and longitude of turning points along its perimeter, or for circular areas by a center point and radius. Figure 11.1, an enlarged excerpt from a 1:250,000-scale chart covering Washington, DC, shows a two-part prohibited area near the city center. Part A of P-56 extends eastward from the Potomac River beyond the Capitol, and includes the Washington Monument, the White House, and other prominent government buildings. Part B is a circle with a half-mile radius centered at the U.S. Naval Observatory. The official residence

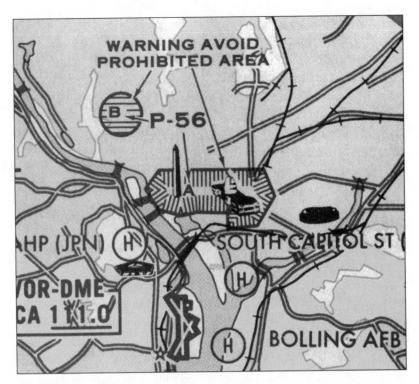

FIGURE 11.1. Prohibited area P-56 as portrayed on FAA Visual Flight Rules (VFR)
Flyway Planning Chart covering Washington, DC.

of the vice president is on Observatory grounds, and the using agency is
the Secret Service, which alone can authorize entry by helicopters or other
aircraft. The prohibition on flying extends upward to an altitude of 18,000
feet, and the chart warns pilots to avoid both areas. About 13 miles south is
area P-73 (not shown), a circle with a half-mile radius centered on Mount
Vernon, a national treasure. The prohibition extends upward only 1,500
feet, and the using agency is the FAA.

By contrast, flying is sometimes allowed within restricted areas, for particular purposes and under certain conditions. More numerous and generally larger than prohibited areas, restricted areas have both a controlling
agency and a using agency as well as numbers, appropriately preceded by
an R. They typically restrict flight near military bases and other defense
facilities. For example, the restriction at area R-4002, around Bloodsworth
Island, Maryland, is in effect "from sunrise to 2400 hours, local time, daily
[and also at] other times as specified in a NOTAM [Notice to Airmen]

issued 48 hours in advance"—to get the latest dope, check with the FAA, which is the controlling agency authorized to act on behalf of the using agency, the Little Creek Naval Amphibious School at Norfolk, Virginia. Flying is also limited within a broad array of Military Operations Areas, Alert Areas, Warning Areas, and National Security Areas—the latter focused on 17 arsenals, weapons plants, and nuclear facilities. Aeronautical charts show their locations and numbers, useful for looking up restricted times and other details.

Few prohibitive maps are as fascinating as the temporary flight restrictions (TFRs) posted on an FAA Web site (tfr.faa.gov) that not only reports duration and justification but also includes maps. What's intriguing is the occasional sudden appearance of a new TFR for a potential catastrophe, safety issue, or diplomatic powwow. In late October 2008, for instance, the Web site listed 29 TRFs. A few marked hazards like a massive forest fire at Big Sur, California, and a menacing volcano at Kilauea, Hawaii—no point in letting pilots interfere with aerial firefighting or encounter a cloud of hot ash. Others pinpointed air shows featuring precision flying teams like the Navy's Blue Angels. And at least two raised questions about the meaning of "temporary." Would the mile-radius circle centered since November 2005 over Vice President Cheney's personal getaway at St. Michaels, Maryland, evaporate at the end of the Bush administration? And what about the three-mile circle set up in March 2003 around the Disneyland Theme Park, in Anaheim, California? Not surprisingly Cheney's "VIP" protection disappeared early in 2009, after he left office, while Disneyland's TFR, attributed to "Special Security Reasons," remained as testimony to the park's preeminence as a cultural icon.

What caught my eye one Thursday afternoon in late October 2008 was a restriction announced the previous day for Saturday morning through early Sunday afternoon near Hagerstown, Maryland. Pop-up maps included a portion of the FAA's 1:500,000-scale aeronautical chart for the area (fig. 11.2, *left*) and a less cluttered "larger map" (fig. 11.2, *right*) showing railroads, major roads, and rivers. The former described the TFR as a circle with a 10-nautical-mile radius centered on prohibited area P-40, described in turn by a three-nautical mile radius around the presidential retreat at Camp David, officially known at the FAA as the "Naval Support Facility, Thurmont." Ah-ha! Though the event was two days away, I could find nothing about it until the weekend, when persistent Web surfing uncovered a report of an overnight presidential visit to the Catoctin Mountain

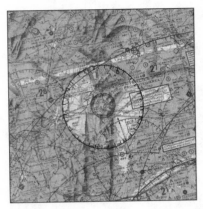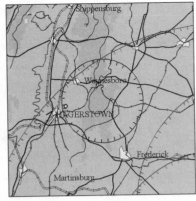

FIGURE 11.2. Maps accompanying the temporary flight restriction announced at the Federal Aviation Administration's TFR Web site on October 22, 2008, for Hagerstown/ Thurmont, Maryland. Left-hand panel (originally in color) shows a substantially reduced and unavoidably cluttered version of the corresponding portion of the FAA's 1:500,000 "sectional" chart; circles with radii of three and 10 nautical miles are centered on Camp David. The more lucid right-hand image, relates the 10-nautical-mile TFR to key geographic features. Area shown on both images is approximately 53 statute miles from left to right.

hideaway. During the Bush administration, TRFs were routinely used, as needed, to temporarily expand prohibited areas over President Bush's Crawford, Texas, ranch and his father's home at Kennebunkport, Maine. Restrictions were slightly less obtrusive under President Obama, whose home on Chicago's South Side was apparently too close to major airports for a standing prohibited area. Ad hoc TFRs were sufficient for brief trips home, perhaps every six or eight weeks.[6] If you want to keep up with the president's or the vice president's domestic travels, check the TFR Web site.

Curious about the history of airspace restrictions on American aeronautical charts, I looked for examples on whatever early charts I could find—not an easy task because map libraries, pressed for storage space, typically discard older editions whenever new ones arrive. I also canvassed the published literature, including government reports, for discussion of the charts' symbolization and content. Although both civilian and military officials were no doubt wary of aircraft passing over firing ranges during artillery practice or inadvertently crashing into oil storage depots or munitions dumps, what I found suggests that formal prohibitions did not appear on American charts until the late 1920s. For example, the illustrated list of "Conventional Signs [for] Aerial Navigation Maps," issued in late 1918 by the War Department (with the endorsement of the U.S. Geographic Board) as the official standard for "all map making departments of the Government," includes symbols for obvious hazards like steel towers and transmission lines but focuses largely on landmarks useful in navigation.[7] A pattern of closely spaced vertical red lines was recommended for marking "Dangerous Landings (Barbed Wire, Gardens, Hedges, Orchards, Quarries, Vineyards, Hop Fields, Ditches, Boulders)," but there was no official symbol to warn pilots away from government buildings or national shrines.

Dramatic increases after World War I in the numbers of aircraft and pilots eventually led to aeronautical charts with standard formats, common symbols, and formally delineated no-fly areas. During the 1920s pilots relied largely on a diverse mix of customized strip charts pioneered by the Army, the Navy, the U.S. Coast and Geodetic Survey, and the Post Office Department as well as private firms like Rand McNally.[8] As the name implies, a strip chart linked a pair of airports a few hundred miles apart. In 1926 Congress passed the Air Commerce Act, which made the Coast and Geodetic Survey responsible for a nationwide series of aeronautical charts. In 1929 the Board of Surveys and Maps, which in 1919 had replaced the Geographic Board as the federal government's arbiter of cartographic standards, recognized the need for an array of 1:500,000 sectional charts covering the entire country, not just narrowly defined flight paths.[9] The previous October the Board had adopted 33 official symbols for aerial navigation, developed jointly by its Committees on Technical Standards and Aerial Navigation Maps.[10] Although some of these symbols had been in use or under consideration as early as 1926, I was unable to find a published precedent for the closely spaced diagonal lines, running from lower left to upper right, and rendered in red to mark a "prohibited area."

Whatever its precedent, the symbol was readily apparent on the 1935 edition of the Washington, DC, sectional chart, where it roped off an extensive area over Chesapeake Bay east of Baltimore, Maryland (fig. 11.3), as part of the Army's Aberdeen Proving Grounds, famed for developing and testing artillery, bombs, and other ordnance. The chart used an identical symbol to outline a similarly extensive area along the lower Potomac, where the Navy had its own firing range.

By contrast, the chart showed no similar restriction for Downtown

FIGURE 11.3. Excerpt from the U.S. Coast and Geodetic Survey's 1935 Washington, DC Sectional Aeronautical Chart, published in color at 1:500,000 and enhanced here to emphasize the prohibited-area symbol and differentiate land from water. The label within the symbol reads, "RESTRICTED AREA. ABERDEEN PROVING GROUNDS AND FIRING RANGE."

Washington, largely because a 1:500,000-scale chart was too generalized for the small area at risk. In 1935 air traffic control was too rudimentary to warrant the more detailed, 1:250,000-scale Flyway Planning Charts and Terminal Charts now available for the country's most congested airports. And as illustrated by the substantially enlarged excerpt of a contemporary Washington-area chart in figure 11.1, even at 1:250,000 there's little room for a detailed treatment.

Delineation of prohibited areas was assigned to the Board on Air-Space Reservations, comprised of the assistant secretaries responsible for aviation matters in the departments of Commerce, War, and the Navy. In 1930 the

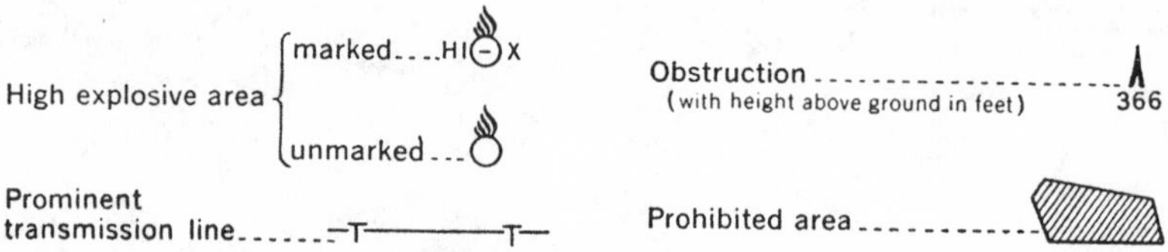

High explosive area
- marked....HI-⊖-x
- unmarked...⊖

Obstruction .... ⋏
(with height above ground in feet) 366

Prominent
transmission line.......—T————T—

Prohibited area ....⬭

FIGURE 11.4. Symbols representing flying hazards or restricted airspace on aeronautical charts in the late 1930s.

Board established two categories of restrictions. Class A reservations, for "national defense or other purposes" but limited to "the minimum consistent with actual requirements," were authorized by the president, following the Board's recommendation.[11] By contrast, Class B reservations, for "certified high-explosive danger areas," could be declared to the secretary of commerce without Board approval or an Executive Order. A facility owner could request certification, which required pilots to avoid the areas "except at a height sufficient to permit of a reasonably safe emergency landing" outside the zone. In general, this meant a minimum altitude of 1,000 feet.

Aeronautical cartographers devised a new chart symbol to show these hazards: a small red circle topped with a flame-like flourish (fig. 11.4). But if officials feared a charted warning might be insufficient, the owner could be required to "display distinctive day and night air markings."[12] As a daytime warning, the area would be ringed by signs with "HI-X" painted in chrome-yellow letters at least 30 feet high but lying horizontally. The nighttime warning required either "high-intensity fixed projectors" mounted on towers at the property's corners and "pointed so as to envelope and outline the area over which flying was restricted," or a 1,000-watt revolving beacon with a red cover glass. Marked danger areas thus became useful landmarks, appropriately distinguished on the chart from their unmarked counterparts.

Franklin Roosevelt's first inauguration was a milestone in restrictive cartography. Officials feared dangerous congestion from airborne sightseers as well as noise that could have marred the new president's radio address—"The only thing we have to fear is fear itself"—intended to reassure a nation in the throes of an economic depression. The March 1, 1933, issue of the *Air Commerce Bulletin,* widely circulated among airport administrators and commercial pilots, announced that "the air space over the District of Columbia is herewith designated as a prohibited area from 9 a.m. to 5 p.m. on March 4, 1933, for public safety purposes."[13] The order, issued by the acting secretary of commerce under authority granted by the Air Commerce Act, threatened violators with "a civil penalty of $500." Use of the city's municipal boundary is a good example of a borrowed-border restriction, based on a preexisting map produced for a wholly different purpose.

Two years later, the April 15, 1935, issue of the *Bulletin* announced a permanent ban on "all aircraft . . . at all times . . . at any altitude" from the air space above Downtown Washington.[14] As publicized by the Commerce Department, the prohibition covered "All that area extending one-quarter of a mile in the horizontal plane beyond the outside limits of that section of the city of Washington, DC, and all the land included within its boundaries, which are marked on the northeast corner by the Union Station, on the southeast corner by the Capitol, on the southwest corner by the Naval Hospital (approximately three-eighths of a mile north of the Lincoln Memorial), and on the northwest corner by the Executive Mansion." Whether to underscore the edict's importance or address readers' need to visualize its extent, the next issue included a black-and-white map that enclosed the forbidden zone with a band of closely spaced diagonal lines faintly similar to the official chart symbol for prohibited areas (fig. 11.5).[15] The less precise map of its contemporary counterpart, area P-56A (fig. 11.1), reveals the ban's expansion southward since the 1930s to encompass the Washington Monument and some newer government buildings south of Independence Avenue. Such restrictions are not uniquely American: one doesn't fly over Buckingham Palace, the Taj Mahal, or the prime minister's residence in almost any developed or lesser developed country you can think of.

Today's airspace restrictions are markedly more complex than 70 years ago. Increased traffic has intensified the threat of collision, especially during peak hours, when too many flights attempt to take off or land at too few runways. In addition to its TFR Web site, the FFA posts a dynamic map of "Special Use Airspace" describing a variety of no-fly zones in effect over the next 24 hours.[16] Controlled airspace also extends hundreds of miles beyond the outer edge of the country's "contiguous zone," a mere 24 nautical miles off the coast. Wary of aircraft approaching their shores, the United States, Canada, Japan, and Australia, among others, have declared Air Defense Identification Zones (ADIZs), within which approaching

Map of Prohibitive Area Over Washington, D. C.

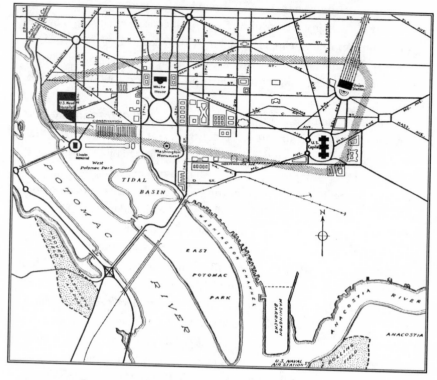

FIGURE 11.5. Permanent flying ban established over downtown Washington, DC, in May 1935.

aircraft must identify themselves and describe their course and destination.[17] Though not explicitly sanctioned by international law, ADIZs are respected by commercial aircraft as a reasonable precaution.

An ADIZ is not just a broad, outwardly focused defense buffer. After the 9/11 attacks on the World Trade Center and the Pentagon by terrorists flying hijacked jetliners, the FAA established the Washington, DC, ADIZ as a 30-nautical-mile circle centered on Reagan National Airport (DCA in fig. 11.6). At its heart is a roughly 15-nautical-mile Flight Restricted Zone, which few aircraft are allowed to enter. Maps publicizing these zones have spared taxpayers the cost of extensive aerial patrols to warn off careless pilots,[18] and surface-to-air missiles are poised to intercept any plane deemed a serious threat.

Cartographically advertised threats to shoot down encroaching aircraft became a humanitarian intervention in 1991, with the proclamation of a

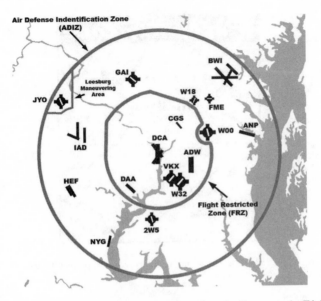

FIGURE 11.6. Following the September 11, 2001, attack on the Pentagon, the FAA defined a pair of concentric zones around Reagan National Airport (DCA) to more rigorously regulate aircraft around Washington, DC.

no-fly zone over northern Iraq as part of a "safe haven" for Iraqi Kurds, who had endured violent repression by Saddam Hussein during the 1980s.[19] An earlier response might have saved countless lives, but the West had little enthusiasm for protecting the Kurds until the Gulf War of 1991, when a coalition of North American and European forces expelled Iraq from oil-rich Kuwait, which Hussein had invaded several months before. Despite their victory, Western forces did not occupy Iraq and thus had limited control over the Republican Guard, Hussein's large army of elite, highly motivated troops who were using poison gas and torture to put down uprisings by Kurds in the north and Shiites in the south. In April 1991, following UN condemnation of the Guard's brutal repression, the United States, Britain, and France declared a "no-fly" zone north of the 36th parallel, to thwart aerial attacks on the Kurds.[20] A southern no-fly zone, declared in August 1992 to protect dissident Shia south of the 32nd parallel, was later expanded to cover areas below the 33rd parallel.[21] AWACS radar planes would monitor compliance, and jet fighters were constantly in the air to repel violators. In linking the "no-fly" edicts to whole-number latitudes, the U.S. Air Force map in figure 11.7 was a stern warning more easily communicated than enforced. Although Iraqi violations often went

FIGURE 11.7. No-fly zones as a cartographic intervention to protect Kurds and Shiia from Iraq's Republican Guard.

unpunished, no-fly zones were used more rigorously and effectively during NATO peacekeeping operations in Bosnia in the mid-1990s.[22]

Don't Go There, Don't Do That

Not all no-go prohibitions are advertised cartographically. Over the years I've heard stories of maps, issued during the Cold War, showing places in the United States that Soviet diplomats and journalists were not allowed to visit. But the only thing I unearthed, after searching databases and querying past and current State Department geographers and cartographers, was a detailed list of forbidden cities, towns, and facilities. Makes sense, though: federal officials could whip up, tweak, and put out a list more quickly (and less expensively) than a detailed map. Let the damned Commies make their own maps—that's how they treated American visitors since May 1941, when Stalin declared large parts of the USSR off-limits.[23] Both countries' decrees reflect the ease of anchoring travel restrictions to an existing canvas of administrative boundaries—another instance of borrowed borders.

The United States had objected to Soviet travel restrictions in 1952, by requiring Russian diplomats or journalists who wanted to venture more than 25 miles beyond the center of New York City or Washington, DC,

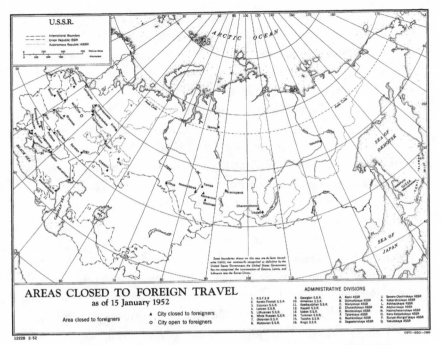

AREAS CLOSED TO FOREIGN TRAVEL
as of 15 January 1952

Area closed to foreigners

▲ City closed to foreigners
○ City open to foreigners

FIGURE 11.8. State Department map published in 1952 shows cities in the U.S.S.R. closed to foreigners.

to clear their plans with the State Department. To justify this restriction, the announcement included a map (fig. 11.8) showing 35 major cities in the USSR, all but two of them (Moscow and Leningrad) closed to outsiders. The forbidden zone was apparently too intricate for a page-size map. According to conservative writer Arnold Beichman, at one time 98 percent of the Soviet Union was closed to foreigners.[24] Obviating a more detailed depiction was the Soviet practice of not admitting the existence of secret cities left off official maps,[25] and the American intelligence community's penchant for not letting on how much it actually knew of Soviet geography.

The Russians had substantially reduced the "closed" area in June 1953, but the United States was not impressed. In January 1955, after nearly two years of dickering with Moscow over the 30 percent of Soviet territory still off limits to Americans, Washington responded with a geographically explicit list of rigid restrictions.[26] "Soviet citizens in possession of U.S.S.R. passports" were now banned from a 15-mile-wide strip along the Canadian and Mexican borders, the shorelines of the Great Lakes, four entire states

(Connecticut, Delaware, Massachusetts, and Rhode Island), 865 counties in 35 other states, and 16 additional cities outside the specifically closed counties. The latter ranged in importance from Atlanta, Georgia, to Johnstown, Pennsylvania. What's more, highway travel to and from New York and Washington was confined to specific routes. Oddly, 64 cities "in otherwise closed areas" were declared "open to travel by Soviet citizens." Thus, while Oneida County, New York, was otherwise out of bounds, Russian travelers could visit its county seat, Utica. And while Camden County, New Jersey, and Philadelphia, immediately across the Delaware River, remained "closed," the city of Camden was officially "open." Perhaps these cities weren't needed to match the Soviets tit for tat.

The accompanying announcement left no doubt that Washington's list was retaliatory. Describing the new regulations as "comparable to those which the Soviet Union has imposed, presumably for reasons of security," the State Department proposed a swift and amicable solution: "Should the Soviet Union conclude that the international situation were such that security requirements enabled it to liberalize its regulations restricting the travel of U.S. citizens in the Soviet Union, the U.S. Government would in turn be disposed to reconsider in the same spirit its own security requirement."

Washington tweaked the list more than once, apparently in response to Russian accommodations. The 1963 roster reported only 754 closed counties in 49 states[27]—Delaware's omission seems an odd oversight—and the 1983 list further reduced the total to 609 counties in 49 states, apparently because the USSR had cut its closures back to 20 percent.[28] While many formerly closed counties had been opened, some new ones were now off limits. The New York Times reporter who analyzed the 1983 list attributed the recent closing of Silicon Valley to "fear of Soviet spying" but noted that "no Russians are known to have been expelled for travel violations." Sometime around the end of the Cold War Russia apparently concluded that travel restrictions, like the arms race, were no longer essential.

However reluctant to make maps telling Russian visitors where they couldn't go, the government relied on cartographic edicts to force thousands of Japanese Americans into "relocation centers" during World War II. On February 19, 1942, roughly two and a half months after Japan's surprise attack on Pearl Harbor, President Roosevelt signed Executive Order 9066: because the war required "every possible protection against espionage and against sabotage to national-defense material," Roosevelt gave Army commanders broad discretion to delineate "military areas . . . from which any and all persons may be excluded."[29] Though the wording did not specifically target Americans of Japanese descent, the intent was clear: send citizens who might be loyal to Emperor Hirohito to austere detention camps far inland, in the mountains or desert. The Army's initial response was the delineation of 135 prohibited areas around airports, military bases, dams, power plants, and other sensitive facilities.[30] So-called exclusion orders required the removal of "all persons of Japanese ancestry, both alien and non-alien," from areas like the zone in figure 11.9, near San Francisco Bay between Berkeley and Oakland. Posted prominently, exclusion maps subtly encouraged fearful residents to report fugitives. This sad and embarrassing episode is further evidence of the power of prohibitive cartography.

A more compassionate cartography, at least from a human standpoint, regulates hunting. Maps are an essential tool for state wildlife conservation agencies that serve sportsmen and women, sporting goods dealers, and tourist businesses by maintaining a robust but nondestructive supply of game animals. Because food supplies, natural predators, and the intensity of hunting vary geographically, even small states like New Hampshire and Vermont are divided into management territories with boundaries published both online and in print. The boundaries appear on a variety of maps because wildlife managers need to vary the length of the hunting season from species to species, place to place, and year to year as well as regulate the weapons a hunter can use and the sex, size, age, and number of animals he or she may "bag."

New York State, which lets hunters "take" deer with a bow or firearm, makes a special provision for muzzle-loading firearms, back-to-basics single-shot rifles or pistols with at least a 0.44-inch bore. As figure 11.10 shows, seasons that vary markedly among the state's 85 Wildlife Management Units (WMUs) in timing and duration require careful attention to WMU boundaries, described in detail both cartographically and verbally on the Department of Environmental Conservation's Web site.[31] In addition to a variety of regular hunting licenses and special muzzle-loading permits, the department also issues Deer Management Permits (DMPs), which allow hunting of antlerless deer, generally adult females, to promote population control. DMPs are issued for specific WMUs, in

PROHIBITED AREA
EXCLUSION ORDER NO. 27
Western Defense Command and Fourth Army

C. E. Order 27

This Map is prepared for the convenience of the public; see the Civilian Exclusion Order for the full and correct description.

FIGURE 11.9. Example of a cartographically illustrated "exclusion order" used during World War II to forcibly relocate Japanese Americans to inland detention camps.

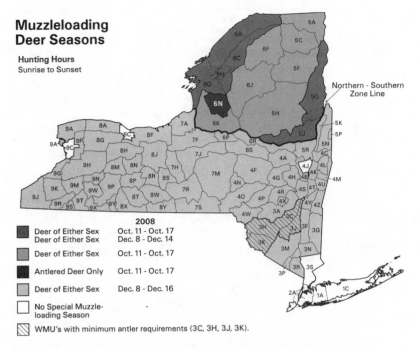

Muzzleloading Deer Seasons

Hunting Hours
Sunrise to Sunset

Northern - Southern
Zone Line

2008

■ Deer of Either Sex   Oct. 11 - Oct. 17
    Deer of Either Sex   Dec. 8 - Dec. 14

■ Deer of Either Sex   Oct. 11 - Oct. 17

■ Antlered Deer Only   Oct. 11 - Oct. 17

□ Deer of Either Sex   Dec. 8 - Dec. 16

□ No Special Muzzle-   -
loading Season

▨ WMU's with minimum antler requirements (3C, 3H, 3J, 3K).

FIGURE 11.10. The New York State Department of Environmental Conservation published its Muzzleloading Deer Season Map in color on the Internet.

limited numbers based on estimates of the current population and the food supply.[32] Compliance requires an appreciation of mapped boundaries.

Because places are often radically dissimilar, every scientifically or politically managed activity is a candidate for prohibitive cartography. Signs might suffice where distances are short or memory needs refreshing—familiar examples include land posted for "no hunting or trespassing" and the biennial ban on electioneering within a hundred feet of polling places. By contrast, maps not only alert the public to spatially intricate pronouncements with grave consequences but also help scientists and administrators assess their own thoroughness. Environmental maps are inherently restrictive because people need to be told not to situate structures on fault lines or in flood zones; bury septic tanks in shallow, highly permeable soils; ravage fragile dunes with all-terrain vehicles; and expect vegetable plants or ornamentals to thrive outside their climatically determined hardiness zones. The needless evacuation of Nisei notwithstanding, most cartographic restrictions are unabashedly beneficial.

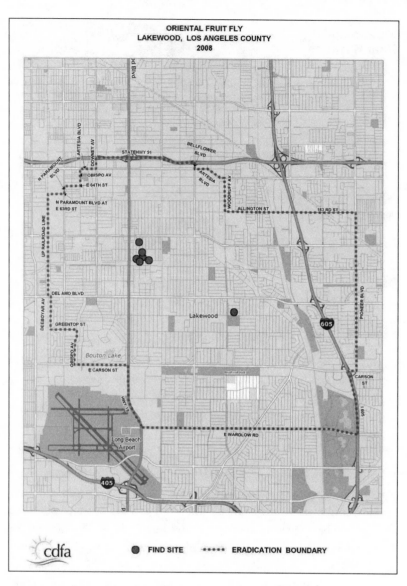

FIGURE II.II. Oriental fruit fly eradication map, Lakewood, California.

Quarantine is a case in point. Agricultural and public health officials have necessarily draconian enforcement powers, as when the California Department of Food and Agriculture must stifle an infestation of Oriental fruit flies. Because this invasive species is a serious threat to 230 types of crops, a map based on the detection of a mere eight adult flies (fig. 11.11) helped the department assert its right and responsibility to conduct an eradication program in a 75-square-mile quarantine area, within which officials removed fruit from trees, embargoed outbound shipments of produce, and begged the public not to take fruit or vegetables beyond the delineated boundary.[33] In many contexts, a prohibitive map is not just an order but a plea for common sense.

CHAPTER
TWELVE

....

ELECTRONIC
BOUNDARIES

Twentieth-century satellite communications triggered a new brand of prohibitive cartography. Developed during the Cold War to keep cruise missiles on a prescribed course, the global positioning system (GPS) joined with wireless telephony and the geographic information system (GIS) to create real-time location tracking. A monitor attached to a delivery van, an Alzheimer's patient, or a level 3 sex offender reports its precise position to a tracking center, where an electronic map that defines a narrow range of acceptable locations can flag a subject who seems lost or suspiciously out-of-bounds. GPS-based monitoring of potentially dangerous travel is the more Orwellian side of the newly emerged location-based services (LBS) industry, known largely for in-vehicle navigation systems.[1]

Location tracking is revolutionary because compliance with prohibitive cartography no longer depends on the vigilant human map-reader who understands and respects graphically coded property lines or administrative boundaries. The effectiveness of geospatial tracking still requires social acceptance of a particular restriction as necessary or appropriate, but the ability to micromanage travel has obviously upped the ante.

Centuries of well-documented abuses of civil liberties by obsessed tyrants, paranoid politicians, and overprotective parents make the potential harm of geospatial tracking easily understood if not chillingly ominous. Particularly frightening is the prospect of instant punitive feedback far out of proportion to an assumed infraction. For example, instead of telling a registered sex offender who's loitering around an elementary school to move on, a "third-generation" tracking system could inject a sleeping potion into his bloodstream—or threaten to do so unless he cleared out immediately.[2] And if geospatially triggered punishment can keep kids safe from perverts, why not stifle known dissidents who stage a seditious protest in front of City Hall or the Federal Building?[3] No less Orwellian is the electronic pinch administered to the curious child who momentarily departs her prescribed route home from school for a close look at a unique flower in a nearby field. That's an example touted by geographer Jerry Dobson, who in 2000 worried about a mindless use of subdermal chips with GPS tracking.[4]

Wary of widening geospatial surveillance, Dobson and fellow GIS expert Peter Fisher coined the term "geoslavery" to describe "a practice in which one entity, the master, coercively or surreptitiously monitors or exerts control over the physical location of another individual, the slave."[5] Images of antebellum plantation society aside, geoslavery has become an accepted practice among employers wary of misuse of company vehicles— whether the worker who disappears for an hour is pursuing an extramarital affair or monopolizing a stool at the donut shop, unauthorized side trips undermine the bottom line. Geospatial shadowing is also gaining traction among auto-insurance and car-rental firms leery of clients who drive recklessly or park in areas with a history of vandalism.[6] Privacy advocates might grumble over creeping acquiescence, but society willingly trades steady employment and lower prices for an electronic leash.

While efforts to stifle goofing off or bad driving are hardly evil, especially when the subject is aware of the surveillance, location tracking is undeniably dangerous when the geomaster is a jealous spouse, an insecure lover, or a psychotic stalker. Dobson and Fisher were especially troubled by a Turkish teenager whose throat was cut after she disgraced her family by attending a movie without permission. Although neither GPS nor GIS nor cellular wireless was implicated in this so-called honor killing, the potential abuse of low-cost person-tracking bracelets packaged with other "mobile computing" services by a global LBS conglomerate is readily apparent. When the penalty for being in the wrong place is death, a vicious beating, or psychological torture, it matters little whether the venue is the Middle East or an American suburb.

Does the capability for abusive applications make these geospatial tools inherently dangerous, or as some critics see them, "value-laden"? Not really, or at least to no greater degree than trucks or plumbing. Put armor and a gun on a truck, and it becomes an instrument for invading neighboring territory and slaughtering helpless civilians. But attach a truck body suitable for hauling, and it can move grain from the field to the mill, flour from mill to bakery, or bread from the bakery to your local grocery. Both tanks and bread trucks are trucks, and thus share such basics as the internal combustion engine, a rigid frame with wheels on axels, and systems for steering and braking. What's more, tanks can be essential for national defense, especially against an enemy with tanks. Similarly ambiguous, plumbing can channel industrial waste into a lake or river or separate our drinking water from fecal matter. Rejecting the "value-laden" hypothesis doesn't deny the need for a wary approach to geospatial surveillance: like any sophisticated technology, it's prone to unintended consequences.

Personal tracking systems are no less ambiguous. Low-cost GPS-enabled ankle bracelets can reduce the number of convicts behind bars, where abuse by other prisoners is common and neophytes too easily learn tricks of the trade from more experienced felons. While prison remains the safest defense against violent or unpredictable convicts, rehabilitation and restitution are more likely outside the penitentiary—assuming government provides appropriate counseling and surveillance. A reliable electronic leash can be less costly than keeping a convict in jail, and even though the threat of prosecution is not a parolee's purest motive for keeping clean, vigilant surveillance is a rational recipe for reducing recidivism. And when that vigilance includes compiling a continuous history of the subject's travels, tracking technology can help the courts better monitor compliance with restraining orders[7] and help the police rule out potential suspects who were nowhere near a crime scene when the incident occurred.

The downside is equally involved. Though an attractive alternative to incarceration, a wireless leash is never wholly reliable: a determined subject can remove or disable the tracking device, and as any driver with a GPS navigation system and a preference for underground parking is well aware, satellite tracking doesn't work indoors.[8] Truth be told, the low-power GPS

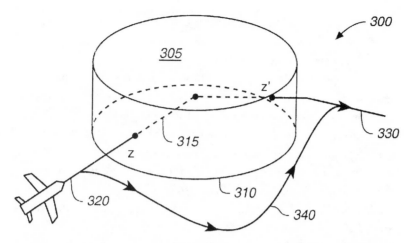

FIGURE 12.1. Collision avoidance system described in figure 3 for U.S. patent 6,675,095 uses numbers to identify key elements. In this example, the "avoidance trajectory" labeled 340 diverts aircraft 320 away from restricted area 305. The numbers 300 and 310 indicate the upper and lower altitudes of the prohibited area, and 315 and 330 identify the initial course and its continuation, which intersect at a turning point within the restricted airspace.

signal is also vulnerable to deliberate or inadvertent jamming.[9] More troubling to civil libertarians is the ease with which society might criminalize a wider range of disagreeable behaviors. Imagine how George Orwell's Big Brother might have used GPS if satellite surveillance were available in the late 1940s. And there's also the lazy parent who uses satellite tracking as a substitute for teaching values and trust.

Criminal justice and irresponsible parenting notwithstanding, geospatial tracking has a bright, less fuzzy side. Among the most promising applications is collision avoidance, particularly important for unmanned aerial vehicles or in congested airspace, where air traffic controllers must communicate with pilots while watching radar screens for planes that are too near the ground, skyscrapers, or each other.[10] GPS-assisted air traffic systems, under development for decades, might eventually let planes land, take off, and otherwise operate as effortlessly and safely as the automated monorail trains at large airports. Details to be worked out include the automated recognition and circumvention of flight restrictions like the cylindrical "no-fly" zone in figure 12.1, from a patent application filed in 2001.[11] While simplistic diagrams make automatic flying look simple, a host of safeguards are needed to assure that a "tamper-resistant" autopilot will recognize and respect permanent and temporary flight restrictions (TFRs) on an electronic aeronautical chart, thereby diminishing the likelihood of false alarms that summon jet fighters.

Other approaches focus on better informing pilots rather than overriding their authority. A NASA study of strategies for reducing inadvertent incursions of TFRs focused on the traditional communication of "security supporting airspace" (SSA) boundaries through Flight Service Station (FSS) briefings and NOTAMs (Notices to Airmen). The typical SSA boundary is a circle identified by its radius and the coordinates of its center, and reported verbally, rather than cartographically. It's up to the pilot whether to plot SSA boundaries on a map. According to human-factors expert Michael Zuschlag, "forcing pilots to manually plot a SSA from a text description provides an opportunity for errors in transcription and interpretation of the NOTAM."[12] What's more, busy pilots often decline to plot SSAs that could become relevant should erratic weather require a change in flight plan. Geospatial technology can promote more careful observance of TFRs by putting a highly relevant, rigorously up-to-date aeronautical chart in the cockpit. Whether displayed in real time on the flight deck or printed before departure, maps like the example in figure 12.2 focus the pilot's attention on key details too easily ignored, misinterpreted, or inaccurately plotted. Also helpful is an automatic map-reading module able to relate the plane's current position to the map and alert the pilot to an imminent intrusion.

Down below, GPS can keep cars, trucks, and trains moving safely. For commuter railways in particular, "positive train control" systems can maintain a safe distance between trains and automatically apply the brakes when an inattentive engineer runs a signal or ignores a speed restriction.[13] While similarly automatic interventions might seem excessive for highway vehicles, geospatial tracking can warn truck drivers of low-clearance hazards like two notorious railroad bridges that cross highways in Syracuse, New York. State transportation officials have tried a variety of warnings— signs displaying the vertical clearance to a tenth of a foot (with a small tolerance deducted for safety, no doubt), yellow diamond-shaped caution signs depicting the top of a truck striking the bottom of a bridge, larger signs, multiple signs, large reflective orange stripes painted along the sides of the bridge, and an active system that triggers flashing lights—but each year the sturdy steel bridges peel the roof off one or two semis.

More generally, a map-savvy navigation system could warn truckers

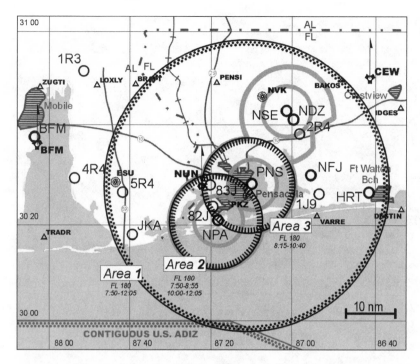

FIGURE 12.2. Hypothetical air navigation map with optimal detail for flight restrictions promotes clarity by suppressing terrain and other details.

about a range of weight, type-of-load, and time-of-day restrictions. And if voluntary compliance fails, their GPS navigators could be programmed to alert highway officials to unwarranted departures from prescribed truck routes. While a tattletale GPS would trigger resentment, and perhaps invite tampering, a system that not only monitors real-time maps of accidents, construction projects, and general congestion but also suggests alternative routes could partly compensate for the implied lack of trust.[14]

Trust and societal acquiescence have always played a key role in prohibitive cartography. People trust maps or at least begrudgingly accept them as necessary tools of governmental administration. Along with their diverse range of restrictive roles, maps have helped validate, or "naturalize," the concepts of private property, the territorial state, congressional redistricting, and land-use zoning. What's on the map is commonly understood as either true or worth fighting over. And where indigenous peoples or local citizens groups challenge the official cartographer's data or assumptions, perhaps by creating "counter maps" as an alternative, their objections con-

FIGURE 12.3. Google Maps image showing marked contrast along the southeastern edge of the Naval Observatory between the pixilated area inside Observatory Circle and the much sharper features outside the perimeter road.

firm the map's hallowed role in restricting activity or movement.[15] We've grown to expect and obey prohibitive maps, and in an era of escalating technological complexity, they've become more common.

Everyday use of electronic mapping tools on the Internet reflects not only the increased pervasiveness of restrictive maps but also the growing democratization of cartography. This promising trend is fragile and too easily reversed. Be wary of public officials who exploit fears of terrorism in denying access to public or quasi-public data and be especially skeptical of small encroachments that make little sense. A case in point is the blurred image of the Naval Observatory on Google Maps (fig. 12.3), apparently because the Bush-Cheney administration feared an attack on the Vice President's Residence, located on Observatory grounds. This ostensibly minor alteration is both frightening and perplexing insofar as a RAND Corporation study commissioned by the National Imaging and Mapping

Agency (now the National Geospatial-Intelligence Agency) after the 9/11 attack on the World Trade Center and the Pentagon concluded that the widely perceived threat of publicly accessible geospatial information was bogus—blurring images or taking down Web sites with publicly funded geospatial data will not thwart terrorists, who can easily glean useful information from other, unrestricted sources.[16] Ironically, Google Maps offers comparatively sharper images of the White House and the Pentagon.[17]

Manipulation of what maps show is far less troubling than manipulation of how they are used. Electronic systems able to relate the position of a person or vehicle to computer-coded boundaries have unprecedented power in restricting where we go, and by extension, what we do. Although the momentum of prohibitive cartography has been subtle and largely unquestioned, its threats to liberty and privacy demand a healthy skepticism of utopian claims for geospatial technologies. Were Orwell still alive, a book titled *CAFO (Confined Animal Feeding Operation)*—a plausible futuristic strategy for coping simultaneously with environmental protection and unconstrained population growth—seems an obvious sequel to his *Animal Farm* and *1984*.

NOTES

Chapter Two

1. Alfred Cornell Mulford, *Boundaries and Landmarks: A Practical Manual* (New York: D. Van Nostrand Co., 1912), 6.
2. Ibid., 17.
3. Ibid., 26.
4. Ibid..
5. Ibid., 40.
6. F. J. Marschner, *Land Use and Its Patterns in the United States,* Department of Agriculture Handbook no. 153 (Washington, DC: Government Printing Office, 1958), 10–21.
7. For the full text, see U.S. Library of Congress, An Ordinance for Ascertaining the Mode of Disposing of Lands in the Western Territory, Documents from the Continental Congress and the Constitutional Convention, 1774–1789, American Memory, http://memory.loc.gov/ammem/collections/continental/index.html.
8. U.S. Bureau of Land Management, *Manual of Instructions for the Survey of the Public Lands of the United States,* 1973 ed. (Washington, DC: Government Printing Office, 1973), 60–61.
9. Norman J. W. Thrower, *Original Survey and Land Subdivision: A Comparative Study of the Form and Effect of Contrasting Cadastral Surveys* (Chicago: Rand McNally, for the Association of American Geographers, 1966).

10. *Land Service Bulletin* 4, no. 8 (1920): 10–12, quoted in Milton Conover, *The General Land Office: Its History, Activities and Organization* (Baltimore: Johns Hopkins Press, 1923), 42.

11. Walter G. Robillard, Donald A. Wilson, and Curtis M. Brown, *Brown's Boundary Control and Legal Principles*, 5th ed. (New York: John Wiley and Sons, 2003), 343–46. The title commemorates surveyor Curtis Brown, the original author, who died in 1993.

12. Ibid., 344.

13. Ibid., 345.

14. Ibid..

15. Walter G. Robillard, Donald A. Wilson, and Curtis M. Brown, *Evidence and Procedures for Boundary Location*, 4th ed. (New York: John Wiley and Sons, 2002), 298.

16. *Enos v. Casey Mountain, Inc.*, 532 So. 2d 703 (Fla. App. 5 Dist. 1988).

17. *Enos v. Casey Mountain, Inc.*, 532 So. 2d 703 at 706.

18. James G. Titus, "Rising Seas, Coastal Erosion, and the Takings Clause: How to Save Wetlands and Beaches without Hurting Property Owners," *Maryland Law Review* 57 (1998): 1279–1399, esp. 1367.

19. Walter Robillard and Lane J. Bouman, *Clark on Surveying and Boundaries*, 7th ed. (Charlottesville, VA: Lexis Law Publishing, 1997), 785 –93.

20. *Peuker v. Canter*, 63 P. 617 (Kans. Sup. Ct. 1901); and Robillard and Bouman, *Clark on Surveying and Boundaries*, 832–33, 864.

21. Robillard, Wilson, and Brown, *Brown's Boundary Control and Legal Principles*, 218, 234–35.

22. James A. Simpson, *River and Lake Boundaries: Surveying Water Boundaries— A Manual* (Kingman, AZ: Plat Key Publishing, 1994), 19, 310–18.

Chapter Three

1. Because this book looks at boundaries across a wide range of map scales, I am wary of the confusion that can occur when discussing countries like the United States of America and political subdivisions like the state of New Jersey. Political theorists, including most political geographers, typically write of international borders as separating "states" (rather than "countries"), which works well enough in most places outside the United States, where "the state" is unambiguously at the top of the political hierarchy and "the province" is the next step down. While the Framers of our Constitution envisioned a loose federation of independent states, each with its own constitution, legislature, and militia, what evolved is a strong federal government that tolerates modest diversity among its 50 provinces, each administered by a "state government" like New Jersey's. To avoid confusion, I use "state" largely for New Jersey and other U.S. provinces. I also steer clear of terms like "sovereign state," which seems redundant, and "nation-state," which recognizes the cohesiveness of

a common language and culture that not all countries enjoy. All the same, "nation" is a convenient synonym for "country," even though many so-called nations are culturally fragmented. What's important is that boundaries between nations (or countries) have cartographies distinct from those at lower levels of the administrative hierarchy, examined in later chapters.

2. Jo Best, "How Eight Pixels Cost Microsoft Millions," *CNET News*.com, August 19, 2004, http://news.com.com/2100-1014_3-5316664.html (accessed August 23, 2007).

3. Raymond Chen, *The Old New Thing: Practical Development throughout the Evolution of Windows* (New York: Addison-Wesley, 2006), 20.

4. Jasper Copping and Melissa Kite, "New EU Map Makes Kent Part of Same 'Nation' as France," *Sunday Telegraph* (London), September 3, 2006.

5. European Commission, Directorate General for Regional Policy, Interreg III: Maps, http://ec.europa.eu/regional_policy/interreg3/carte/cartes_en.htm (accessed December 28, 2006).

6. Copping and Kite, "New EU Map Makes Kent Part of Same 'Nation' as France."

7. Map posted at http://img.dailymail.co.uk/i/pix/2006/09/mapeu040906_228x322.jpg to accompany Tim Shipman, "New Map of Britain That Makes Kent Part of France . . . and It's a German Idea," *Daily Mail* (London), September 4, 2006 (accessed December 26, 2006).

8. Jean-Noël Marchandiau, *L'Illustration: 1843–1944: Vie et mort d'un journal* (Toulouse: Éditions Privat, 1987), 300–302

9. P. J. Philip, "1919 French Map Cause of Furor; Printer Drew in Disputed Borders," *New York Times*, April 6, 1940.

10. Jeffrey S. Wilson, Timothy S. Brothers, and Eugenio J. Marcano, "Remote Sensing of Spatial and Temporal Vegetation Dynamics in Hispaniola: A Comparison of Haiti and the Dominican Republic," *Geocarto International* 16 (June 2001): 5–16.

11. Tom O'Neill, "DMZ: Korea's Dangerous Divide," *National Geographic* 204 (July 2003): 2–27.

12. U.S. Central Intelligence Agency, "The Disputed Area of Kashmir," map 76027AI (R00744), June 2002.

13. U.S. Department of State, "Background Note: Saudi Arabia," http://www.state.gov/r/pa/ei/bgn/3584.htm (accessed September 24, 2007).

14. International Boundary Commission, http://www.internationalboundary commission.org/ (accessed September 3, 2007).

15. Colin McDonald, "Landscaping Wall Turns into Political, Legal Mess," *Seattle Post-Intelligencer,* July 18, 2007; and Tomas Alex Tizon, "Mrs. Leu, Tear Down That Wall," *Los Angeles Times,* May 26, 2007.

16. Eunice Moscoso, "Lawmaker Proposes Barricade on Border," *Atlanta Journal-Constitution,* November 10, 2005.

17. Alicia A. Caldwell, "Border Barrier Accidentally Built on Mexican Soil," Associated Press, June 29, 2007.

18. See, for example, Tom Baldwin, "Fences to Keep Out Wetbacks Can't Be Built without Them," *The Australian*, March 1, 2006.

19. The international boundary also includes a 24-mile (37-kilometer) stretch of the Colorado River. International Boundary and Water Commission, http://www.ibwc.state.gov/ (accessed September 7, 2007).

20. "Convention between the United States of America and the United States of Mexico Touching the International Boundary Line Where It Follows the Bed of the Colorado," November 12, 1884, http://www.ibwc.state.gov/Files/TREATY_OF_1884.pdf (accessed September 7, 2007).

21. James E. Hill, Jr., "El Chamizal: A Century-old Boundary Dispute," *Geographical Review* 55 (1965): 510–22.

22. Larman C. Wilson, "The Settlement of Boundary Disputes: Mexico, the United States, and the International Boundary Commission," *International and Comparative Law Quarterly* 29 (1980): 38–53.

23. "Convention between the United States of America and the United Mexican States for the Solution of the Problem of the Chamizal," *International Legal Materials* 2 (1963): 874–81.

24. Quoted in S. Whittemore Boggs, *International Boundaries: A Study in Boundary Functions and Problems* (New York: Columbia University Press, 1940), 86.

25. Thomas H. Holdich, *Political Frontiers and Boundary Making* (London: Macmillan and Co., 1916), x.

26. Ibid., 34.

27. Ibid., 46.

28. Ibid., 162.

29. Ibid., 184.

30. Julia Lovell, *The Great Wall: China against the World, 1000 BC—AD 2000* (New York: Grove Press, 2006).

31. Colin Martin, "Hadrian's Wall," *British Heritage* 24 (September 2003): 26–35.

32. Abigail Cutler, "Security Fences," *Atlantic Monthly* 295 (March 2005): 40.

33. Shaul E. Cohen, "Israel's West Bank Barrier: An Impediment to Peace?" *Geographical Review* 96 (2006): 682–95. The barrier's route has been revised and altered several times; for maps of plans, revisions, and completed sections, see Israel Ministry of Defense, Israel Security Fence, www.securityfence.mod.gov.il (accessed September 20, 2007).

34. Mark Monmonier, *From Squaw Tit to Whorehouse Meadow: How Maps Name, Claim, and Enflame* (Chicago: University of Chicago Press, 2006), 112–21.

35. Holdich, *Political Frontiers and Boundary Making*, 97.

36. John O'Loughlin and Herman van der Wusten, "Political Geography of Panregions," *Geographical Review* 80 (1990): 1–20.

37. Edward Dommen, "What Is a Microstate?" in *States, Microstates, and Islands*, ed. Edward Dommen and Philippe Hein, 1–15 (Beckenham, Kent, UK: Croom Helm, 1985).

38. Yaroslav Trofimov, "SMOM Is a Mouse That Roars for Respect as a Bona

Fide Nation—Recognized by 87 Countries, Leaders of Knights of Malta Have No Citizens or Territory," *Wall Street Journal,* June 28, 2001.

39. See Order of Malta, http://www.orderofmalta.org/ (accessed October 1, 2007); and Matt Rosenberg, "The Sovereign Military Order of Malta Is Not a Country," http://geography.about.com/od/politicalgeography/a/knights malta.htm (accessed October 1, 2007).

40. Joseph T. Glatthaar and James Kirby Martin, *Forgotten Allies: The Oneida Indians and the American Revolution* (New York: Hill and Wang, 2006), 314.

41. Jack Campisi, "The Oneida Treaty Period, 1783–1838," in *The Oneida Indian Experience: Two Perspectives,* ed. Jack Campisi and Laurence M. Hauptman, 48–64 (Syracuse, NY: Syracuse University Press, 1988).

42. U.S. Geological Survey, Oneida, NY, 15-minute topographic map (1895), 1:62,500

43. C. H. Birdseye, ed., *Topographic Instructions of the United States Geological Survey,* U.S. Geological Survey Bulletin 788 (Washington, DC: U.S. Government Printing Office, 1928), plate 20.

44. Ibid., 236.

45. U.S. Geological Survey, Tully, NY, 15-minute topographic map (1900), 1:62,500.

46. Birdseye, *Topographic Instructions,* 327.

47. *Oneida Indian Nation v. County of Oneida,* 414 U.S. 661 (1974).

48. *County of Oneida v. Oneida Indian Nation,* 470 U.S. 226 (1985).

49. *City of Sherrill v. Oneida Indian Nation of N. Y.,* 544 U.S. 197 (2005).

50. Bureau of Indian Affairs, *Draft Environmental Impact Statement, Oneida Nation of New York Conveyance of Lands into Trust, November, 2006,* ES-17, http://www.oneidanationtrust.net/ (accessed October 5, 2007).

Chapter Four

1. See, for example, Charles Edward Nowell, "The Treaty of Tordesillas and the Diplomatic Background of American History," in *Greater America: Essays in Honor of Herbert Eugene Bolton,* ed. Adele Ogden and Engel Stuiter, 1–18 (Berkeley and Los Angeles: University of California Press, 1945).

2. Alison Sandman, "Spanish Nautical Cartography in the Renaissance," in *Cartography in the European Renaissance,* vol. 3 of *The History of Cartography,* ed. David Woodward, 1095–1142 (Chicago: University of Chicago Press, 2007), esp. 1108–9.

3. Maria Fernanda Alegria and others, "Portuguese Cartography in the Renaissance," in *Cartography in the European Renaissance,* vol. 3 of *The History of Cartography,* ed. David Woodward, 975–1068 (Chicago: University of Chicago Press, 2007), esp. 993–94.

4. The 17-degree estimate was a compromise between Spanish calculations that placed the antimeridian 3 degrees to the west of the islands and Por-

tuguese calculations that situated it 43 degrees to the east. Peter Whitfield, *New Found Lands: Maps in the History of Exploration* (New York: Routledge, 1998), 94. Also see Sandman, "Spanish Nautical Cartography in the Renaissance," 1111–16.

5. M. J. Blakemore and J. B. Harley, *Concepts in the History of Cartography: A Review and Perspective* [Cartographica Monograph 26] (Toronto: University of Toronto Press, 1980), 92.

6. J. D. Hargreaves, "Towards a History of the Partition of Africa," *Journal of African History* 1 (1960): 97–109.

7. Peter Collier, "Imperial Boundary Making in the 19th Century," *Society of Cartographers Bulletin* 38, no. 2 (2004): 45–47.

8. E. H. Hills, "The Geography of International Frontiers," *Geographical Journal* 28 (1906): 145–55; quotations on 150 and 146.

9. Ibid., 147.

10. Collier, "Imperial Boundary Making in the 19th Century."

11. For insights on the role of boundaries in internal and international conflict in Africa and elsewhere, see Alberto Alesina, William Easterly, and Janina Matuszeski, "Artificial States," Harvard Institute of Economic Research, Discussion Paper no. 2115, March 2006, http://www.ssrn.com/; and Stephen A. Kocs, "Territorial Disputes and Interstate War, 1945–1987," *Journal of Politics* 57 (1995): 159–75.

12. Quoted in David Smith, "Dreams of a United Africa," National Geographic News, September 21, 2000, http://news.nationalgeographic.com/news/2000/09/0921_africa.html (accessed December 2, 2007).

13. Michael Heffernan, "The Politics of the Map in the Early Twentieth Century," *Cartography and Geographic Information Science* 29 (2002): 207–26.

14. Geoffrey Martin, *The Life and Thought of Isaiah Bowman* (Hamden, CT: Archon Books, 1980), 81–97; and Neil Smith, *American Empire: Roosevelt's Geographer and the Prelude to Globalization* (Berkeley: University of California Press, 2003), 113–38.

15. Isaiah Bowman, *The New World: Problems in Political Geography* (Yonkers-on-Hudson, NY: World Book Company, 1922), map and quotation on 4.

16. Ellen Churchill Semple, "Geographical Boundaries—II," *Bulletin of the American Geographical Society* 39 (1907): 449–63; quotation on 451.

17. Bowman, *The New World*, 3.

18. Charles Seymour, "The End of an Empire: Remnants of Austria-Hungary," in *What Really Happened at Paris*, ed. Edward Mandell House and Charles Seymour (New York: Charles Scribner's Sons, 1921), 87–111; esp. 105.

19. Jeremy W. Crampton, "The Cartographic Calculation of Space: Race Mapping and the Balkans at the Paris Peace Conference of 1919," *Social and Cultural Geography* 7 (2006): 731–52.

20. Jovan Cvijić, "The Zones of Civilization of the Balkan Peninsula," *Geographical Review* 5 (1918): 470–82; quotation on 470.

21. See, for example, Edward P. Joseph and Michael E. O'Hanlon, *The Case for*

Soft Partition, Saban Center for Middle East Policy Analysis Paper no. 12 (Washington, DC: Brookings Institution, 2007). For a critical assessment of partition's effectiveness, see Nicholas Sambanis, "Partition as a Solution to Ethnic War: An Empirical Critique of the Theoretical Literature," *World Politics* 52 (2000): 437–83.

22. Robert E. Wilson, "National Interests and Claims in the Antarctic," *Arctic* 17 (1964): 15–31.

23. Frank C. Alexander, Jr., "Legal Aspects: Exploitation of Antarctic Resources," *University of Miami Law Review* 33 (1978): 371–423. Because Argentina has made little effort to follow up on its 1927 claim, numerous sources provide a later date for its Antarctic claim; see, for example, Harm J. de Blij, "A Regional Geography of Antarctica and the Southern Ocean," *University of Miami Law Review* 33 (1978): 299–314.

24. Christopher C. Joyner, *Antarctica and the Law of the Sea* (Dordrecht: Martinus Nijhoff, 1992), 42–49.

25. Carlos Escudé, "Argentine Territorial Nationalism," *Journal of Latin American Studies* 20 (1988): 139–65.

26. J. S. Reeves, "Antarctic Sectors," *American Journal of International Law* 22 (1939): 519–21; quotation on 520.

27. Joyner, *Antarctica and the Law of the Sea*, 47.

28. Cornelia Lüdecke, "In geheimer Mission zur Antarktis: Die dritte Deutsche Antarktische Expedition 1938/39 und der Plan einer territorialen Festsetzung zur Sicherung des Walfangs," *Deutsches Schiffahrtsarchiv* 26 (2003): 75–100; and Wilson, "National Interests and Claims in the Antarctic," 21–22.

29. Alexander, "Legal Aspects: Exploitation of Antarctic Resources," 375.

30. Klaus J. Dodds, "South Africa and the Antarctic, 1920–1960," *Polar Record* 32 (1996): 25–42.

31. Martin Ira Glassner, "The View from the Near North—South Americans View Antarctica and the Southern Ocean Geopolitically," *Political Geography Quarterly* 4 (1985): 329–42.

32. Central Intelligence Agency, Antarctica: Research Stations and Territorial Claims, map no. 705786 (543391), September 1985.

33. For the full text of the treaty, see the Antarctic Treaty Secretariat Web site, http://www.ats.aq/. The Consultative Parties consist of the 12 original signatories and 16 countries admitted to consultative status after the mid-1970s. An additional 18 countries have signed the treaty and can attend the meetings as nonvoting makers.

34. For research station data, see Central Intelligence Agency, "Antarctica," World Factbook, https://www.cia.gov/library/publications/the-world-fact book/geos/ay.html. For a recent map, consult https://www.cia.gov/library/publications/the-world-factbook/reference_maps/antarctic.html.

35. Klaus K. Dodds, "Post-Colonial Antarctica: An Emerging Engagement," *Polar Record* 42 (2006): 59–70; and Martin Ira Glassner, *Neptune's Domain: A Political Geography of the Sea* (Boston: Unwin Hyman, 1990), 100–105.

Chapter Five

1. For background, see U.S. Department of State, "United States Responses to Excessive National Maritime Claims," *Limits in the Seas,* no. 112 (March 9, 1992), http://www.state.gov/documents/organization/58381.pdf.
2. Jon M. Van Dyke, "The Republic of Korea's Maritime Boundaries," *International Journal of Marine and Coastal Law* 18 (2003): 509–40.
3. Lewis M. Alexander, *Offshore Geography of Northwestern Europe: The Political and Economic Problems of Delimitation and Control* (Chicago: Rand McNally, for the Association of American Geographers, 1963), table 4 on 60.
4. D. P. O'Connell, *The International Law of the Sea: Volume I* (Oxford: Clarendon Press, 1982), 447–58.
5. United Nations, Convention on the Territorial Sea and the Contiguous Zone, http://untreaty.un.org/ENGLISH/bible/englishinternetbible/partI/chapterXXI/treaty1.asp (accessed January 8, 2008).
6. United Nations, Table of Claims to Maritime Jurisdiction (as at 24 October 2007), http://www.un.org/Depts/los/LEGISLATIONANDTREATIES/PDFFILES/table_summary_of_claims.pdf (accessed January 10, 2008).
7. Ronald Reagan, "Proclamation 5928 of December 27, 1988: Territorial Sea of the United States of America," *Federal Register* 54 (January 9, 1989): 777; and Andrew Rosenthal, "Reagan Extends Territorial Waters to 12 Miles," *New York Times,* December 29, 1988.
8. William J. Clinton, "Proclamation 7209 of August 2, 1999: Contiguous Zone of the United States," *Federal Register* 64 (September 8, 1999): 48701–2; and Philip Shenon, "U.S. Doubles Offshore Zone under Its Law," *New York Times,* September 3, 1999.
9. National Research Council, Committee on Fisheries, *Improving the Management of U.S. Marine Fisheries* (Washington, DC: National Academies Press, 1994), 13–15.
10. Ronald Reagan, "Proclamation 5030 of March 10, 1983: Exclusive Economic Zone of the United States of America," *Federal Register* 48 (March 14, 1983): 10605–6.
11. Bernard R. Nossiter, "Law Signed by 117 Nations; U.S. Opposes It," *New York Times,* December 11, 1982.
12. James L. Malone, "The United States and the Law of the Sea after UNCLOS III," *Law and Contemporary Problems* 46, no. 2 (Spring 1983): 29–36; quotations on 32, 31, and 29.
13. UN Convention on the Law of the Sea, Article 76: Definition of the Continental Shelf, http://www.un.org/Depts/los/convention_agreements/texts/unclos/part6.htm.
14. Philomène A. Verlaan, "New Seafloor Mapping Technology and Article 76 of the 1982 United Nations Convention on the Law of the Sea," *Marine Policy* 21 (1997): 425–34.
15. Ted L. McDorman, "The Role of the Commission on the Limits of the

Continental Shelf: A Technical Body in a Political World," *International Journal of Marine and Coastal Law* 17 (2002): 301–24. For further information on the Commission's responsibilities and procedures, see UN Commission on the Limits of the Continental Shelf, http://www.un.org/Depts/los/clcs_new/clcs_home.htm.

16. UN Commission of the Limits of the Continental Shelf, Submissions to the Commission: Submission by the Russian Federation, http://www.un.org/Depts/los/clcs_new/submissions_files/submission_rus.htm.

17. Quoted in "U.S. Reaction to Russian Continental Shelf Claim," *American Journal of International Law* 96 (2002): 969–70; quotations on 970.

18. C. J. Chivers, "Eyeing Future Wealth, Russians Plant the Flag on the Arctic Seabed, below the Polar Cap," *New York Times,* August 3, 2007.

19. Alex Oude Elferink, "The 1990 USSR-USA Maritime Boundary Agreement," *International Journal of Estuarine and Coastal Law* 6 (1991): 41–52; appendix includes full text of the treaty.

20. Donat Pharand, "Delimitation Problems of Canada (Second Part)," in *The Continental Shelf and the Exclusive Economic Zone: Delimitation and Legal Regime,* ed. Donat Pharand and Umberto Leanza, 171–80 (Dordrecht: Martinus Nijhoff, 1993); and David L. Vanderzwaag and Cynthia Lamson, "Ocean Development and Management in the Arctic: Issues in American and Canadian Relations," *Arctic* 39 (1986): 327–37.

21. Overlapping U.S. and Canadian claims are among the disputed boundaries included on the map "Maritime Jurisdiction and Boundaries in the Arctic Region," issued in July 2008 by the International Boundaries Research Unit at the University of Durham, in Britain. The map and its accompanying briefing notes, online at www.durham.ac.uk/ibru, purport to show "theoretical maximum claims assuming that none of the states claims continental shelf beyond median lines with neighboring states where maritime boundaries have not been agreed to." Also see Randy Boswell, "British Map Tries to Sort Out Conflicting Arctic Claims," *Vancouver Sun,* August 6, 2008.

22. Michael Byers, "We Can Settle This; Let's Trade Oil for Fish," *Globe and Mail* (Toronto), March 11, 2005.

23. UN Convention on the Law of the Sea, Article 121: Regime of Islands, http://www.un.org/Depts/los/convention_agreements/texts/unclos/part8.htm.

24. Pew Charitable Trust's Sea around Us Project, http://www.seaaroundus.org/eez/eez.aspx (accessed January 21, 2008).

25. Nuno Sérgio Marques Antunes, "The Importance of the Tidal Datum in the Definition of Maritime Limits and Boundaries," *International Boundaries Research Unit Maritime Briefing* 2, no. 7 (2000).

26. For an overview, see International Hydrographic Organization, Advisory Board on the Law of the Sea, *A Manual on Technical Aspects of the United Nations Convention on the Law of the Sea—1982,* Special Publication 51, 4th ed. (Monaco: International Hydrographic Bureau, 2006).

27. U.S. Department of State, Bureau of Intelligence and Research, Office of the Geographer, "Maritime Boundary: United States—Venezuela," *Limits in the Seas,* no. 91 (December 16, 1980), http://www.state.gov/documents/organization/58824.pdf.

28. U.S. Department of State, Bureau of Intelligence and Research, Office of the Geographer, "Maritime Boundary: United States—Cook Islands and United States—New Zealand (Tokelau)," *Limits in the Seas,* no. 100 (December 30, 1983), http://www.state.gov/documents/organization/58566.pdf.

29. Ibid.

30. Quotation from Michael Field, "Pacific Mouse Roars for Piece of US," *Dominion Post* (Wellington, New Zealand), February 4, 2006. Also see Ian Parker, "Birth of a Nation? A Speck in the South Pacific Ponders Independence," *New Yorker* 82 (May 1, 2006): 66–75.

31. U.S. Central Intelligence Agency, The World Factbook: United States, https://www.cia.gov/library/publications/the-world-factbook/geos/us.html (accessed January 25, 2008).

32. U.S. Central Intelligence Agency, The World Factbook: Tokelau, https://www.cia.gov/library/publications/the-world-factbook/geos/tl.html.

33. Intergovernmental Panel on Climate Change, *Climate Change 2007: The Physical Science Basis—Summary for Policymakers,* Geneva, Switzerland.

34. Jon Barnett and W. Neil Adger, "Climate Dangers and Atoll Countries," *Climatic Change* 61 (2003): 321–37, graph on 324; and U.S. Central Intelligence Agency, The World Factbook: The Maldives, https://www.cia.gov/library/publications/the-world-factbook/geos/mv.html.

35. Uli Schmetzer, "Archipelago Wages Fight to Stay Afloat," *Chicago Tribune,* December 27, 1999.

36. UN Convention on the Law of the Sea, Article 60: Artificial Islands, Installations and Structures in the Exclusive Economic Zone, http://www.un.org/Depts/los/convention_agreements/texts/unclos/part5.htm.

37. UN Convention on the Law of the Sea, Article 121: Regime of Islands, http://www.un.org/Depts/los/convention_agreements/texts/unclos/part8.htm.

38. Jon Barnett, "Titanic States? Impacts and Responses to Climate Change in the Pacific Islands," *Journal of International Affairs* 59 (2005): 203–19; and Barnett and Adger, "Climate Dangers and Atoll Countries," footnote on 327.

39. For information on U.S. practices, see the National Marine Fisheries Service Web site, at http://www.nmfs.noaa.gov/.

40. Jon M. Ban Dyke, "The Disappearing Right to Navigational Freedom in the Exclusive Economic Zone," *Marine Policy* 29 (2005): 107–21.

41. Mark J. Valencia and Kazumine Akimoto, "Guidelines for Navigation and Overflight in the Exclusive Economic Zone," *Marine Policy* 30 (2006): 704–11.

42. Sarah Lyall, Michael Slackman, and Edward Wong, "Iran Sets Free 15 Britons Seized at Sea in March," *New York Times*, April 5, 2007.

43. International Boundaries Research Unit, University of Durham, Notes on the Iran-Iraq Maritime Boundary, http://www.dur.ac.uk/ibru/resources/iran-iraq/ (accessed March 31, 2007).

44. Anthony F. Shelley, "Law of the Sea: Delimitation of the Gulf of Maine," *Harvard International Law Journal* 26 (1985): 646–54.

45. Federal Geographic Data Committee, Marine Boundary Working Group, *Marine Managed Areas: Best Practices for Boundary Making* (Charleston, SC: NOAA Coastal Service Center, 2006); online at http://www.csc.noaa.gov/products/mb_handbook/.

46. Both terms are defined as "any area of the marine environment that has been reserved by federal, state, territorial, tribal, or local laws or regulations to provide lasting protection for part or all of the natural and cultural resources therein." See the glossary section of the Marine Protected Areas of the United States Web site, http://www.mpa.gov/glossary.html.

47. Amanda Johnson, NOAA Fisheries Service, Northeast Regional Office, telephone conversation, February 6, 2008.

48. NOAA Fisheries Service, Harbor Porpoise Take Reduction Plan, http://www.nero.noaa.gov/prot_res/porptrp/outr.html (accessed February 6, 2008).

49. Piscataqua River at Portsmouth Naval Shipyard, Kittery, Maine—Restricted Areas, *Code of Federal Regulations*, title 33, sec. 334.50.

50. Kittery, Maine—Regulated Navigation Area, *Code of Federal Regulations*, title 33, sec. 165.101.

Chapter Six

1. Morris M. Thompson, *Maps for America: Cartographic Products of the U.S. Geological Survey and Others* (Washington, DC: Government Printing Office, 1979), 80–87.

2. Mei-Ling Hsu, "The Han Maps and Early Chinese Cartography," *Annals of the Association of American Geographers* 68 (1978): 45–60; O. A. W. Dilke, "Maps in the Service of the State: Roman Cartography to the End of the Augustan Era," in *Cartography in Prehistoric, Ancient, and Medieval Europe and the Mediterranean*, vol. 1, *The History of Cartography*, ed. J. B. Harley and David Woodward, 201–11 (Chicago: University of Chicago Press, 1987); and Peter Barber, "Mapmaking in England, ca. 1470–1650," in *Cartography in the European Renaissance*, vol. 3, *The History of Cartography*, ed. David Woodward, 1589–1669 (Chicago: University of Chicago Press, 2007).

3. U.S. Federal Geographic Data Committee, *Development of a National Digital Geospatial Data Framework* (Reston, VA: Federal Geographic Data Committee Secretariat, 1995).

4. See "County Subdivisions," chapter 8 in U.S. Census Bureau, *Geographic Areas Reference Manual*, rev. September 2005, http://www.census.gov/geo/www/garm.html (accessed February 27, 2008).

5. Categories on the map are based on figure 8-1 and table 8-3 in ibid.

6. Lorna Colquhoun, "No People, Plenty of Ballots," *Manchester Union Leader*, November 8, 2006; and New Hampshire Office of Energy and Planning, "2006 Population Estimates of New Hampshire Cities and Towns," July 2007, http://www.nh.gov/oep/programs/DataCenter/Population/documents/pub06.pdf.

7. Two classic studies are John C. Hudson, "North American Origins of Middlewestern Frontier Populations," *Annals of the Association of American Geographers* 78 (1988): 395–413; and John Leighly, "Town Names of Colonial New England in the West," *Annals of the Association of American Geographers* 68 (1978): 233–48.

8. U.S. Census Bureau, *Popularly Elected Officials*, Vol. 1, no. 2 for the Census of Governments, 1992 (Washington, DC, 1995), table 18, p. 20.

9. Jered B. Carr and Richard C. Feiock, "State Annexation 'Constraints' and the Frequency of Municipal Annexation," *Political Research Quarterly* 54 (2001): 459–70.

10. Maps prepared for the 1940 Census revealed 27 other "enclave municipalities." See Richard G. Spencer, "29 Cities within Cities," *National Municipal Review* 37 (1948): 256–58.

11. Charles S. Aiken, "Race as a Factor in Municipal Underbounding," *Annals of the Association of American Geographers* 77 (1987): 564–79; and Daniel T. Lichter and others, "Municipal Underbounding: Annexation and Racial Exclusion in Small Southern Towns," *Rural Sociology* 72 (2007): 47–68.

12. Jo Desha Lucas, "Dragon in the Thicket: A Perusal of Gomillion v. Lightfoot," *Supreme Court Review*, 1961, 194–244; quotation on 195.

13. *Gomillion v. Lightfoot*, 364 U.S. 339, 364 (1960).

14. John Herbers, "Tuskegee Thriving under Biracial Rule," *New York Times*, November 25, 1964.

15. Because unincorporated land is scarce or nonexistent, New Jersey, Pennsylvania, and all New England states but Maine lack traditional annexation statutes. Even so, their laws do allow for amalgamation of towns as well as boundary changes. Paula E. Steinbauer and others, *An Assessment of Municipal Annexation in Georgia and the United States: A Search for Policy Guidance* (Athens: University of Georgia, Carl Vinson Institute of Government, 2002), 15.

16. Kristen M. O'Connor, "Losing Ground: *Seminole* and the Annexation Power of Municipalities in Oklahoma," *Oklahoma Law Review* 58 (2005): 527–49; quotation on 528.

17. Alison Yurko, "A Practical Perspective about Annexation in Florida," *Stetson Law Review* 25 (1996): 699–719; quotation on 702.

18. O'Connor, "Losing Ground."

19. *City of Tuskegee v. Lacey,* 486 So. 2d 393 (Ala. 1985).

20. Ibid., 395.

21. *City of Fultondale v. City of Birmingham,* 507 So. 2d 489, 490 (Ala. 1987).

22. Steinbauer and others, *An Assessment of Municipal Annexation,* 21–22.

23. Ibid., 19.

24. Measurements are based on 1993 editions of U.S. Geological Survey's Elmhurst, Illinois, and River Forest, Illinois, 7.5-minute topographic quadrangle maps, 1:24,000. For a brief history of the airport, see Amanda Seligman, "O'Hare," Encyclopedia of Chicago, http://www.encyclopedia.chicagohistory .org/pages/924.html (accessed March 13, 2008).

25. Based on tabulations of historical annexation activity in Steinbauer and others, *An Assessment of Municipal Annexation,* 66–67.

26. John H. Long, telephone conversation, March 14, 2008. Also see John H. Long, "Atlas of Historical County Boundaries," *Journal of American History* 81 (1995): 1859–63. A massive multivolume work, the *Atlas of Historical County Boundaries* maps every merger, split, or reconfiguration since colonial times. Delving into the state laws that made revised boundaries official and described their positions verbally, Long and his assistants advanced coverage region by region, state by state—a painstaking, often tedious task that began in the late 1970s.

27. William Dollarhide, "The Old Boundary Line Blues," *Genealogy Bulletin* 15 (January/February 1999): 1–12.

28. John H. Long and Kathryn Ford Thorne, *Atlas of Historical County Boundaries: New York* (New York: Simon and Schuster, 1993), 126–29.

29. Ibid., 16.

30. See, for example, John Borchert's theretofore unpublished map of "third-order nodal regions" in Ronald Abler, John S. Adams, and Peter Gould, *Spatial Organization: The Geographer's View of the World* (Englewood Cliffs, NJ: Prentice-Hall, 1971), 568.

31. G. Etzel Pearcy, *A Thirty-Eight State U.S.A.* (Fullerton, CA: Plycon Press, 1973).

32. Ibid., 3.

33. "Boston, Plym., and Boise, Bitt.," *Time* 102 (December 17, 1973): 10.

34. Mark Stein, *How the States Got Their Shapes* (New York: HarperCollins, 2008).

35. Charter quoted in W. C. Hodgkins, "A Historical Account of the Boundary Line between Pennsylvania and Delaware," Appendix 8 in *Report of the Superintendent of the Coast and Geodetic Survey for the Fiscal Year Ending June 30, 1893, Part II—Appendices* (Washington, DC: Government Printing Office, 1895), 177–222; quotation on 184.

36. Lynn Perry, "The Circular Boundary of Delaware," *Civil Engineering* 4 (November 1934): 576–80.

37. The bearing of the longer section is N 3°36'06" W.
38. Rick Hampson, "High Court to Settle Fight for the Delaware," *USA Today*, November 15, 2007; and Susan Warner, "Abundant Energy or Floating Bombs?" *New York Times*, December 4, 2005.
39. Linda Greenhouse, "Court Blocks Plans for New Gas Plant in New Jersey," *New York Times*, April 1, 2008; and *New Jersey v. Delaware*, 552 U. S. ____ (2008).
40. Stein, *How the States Got Their Shapes*, 188–91.
41. *New Jersey v. New York*, 523 U.S. 767 (1998).
42. For a map and boundary description, see *New Jersey v. New York*, 526 U.S. 589 (1999).
43. U.S. Government Accountability Office, *Rural Housing: Changing the Definition of Rural Could Improve Eligibility Determinations*, Report GAO-05-110, December 2004, 17.
44. Glenn Coin, "Oneida Ruling Faces a Test," *Syracuse Post-Standard*, February 23, 2008; and U.S. Department of the Interior, Office of the Assistant Secretary for Indian Affairs, Fact Sheet concerning the Final Environmental Impact Statement for the Oneida Indian Nation of New York's Land-into-Trust Application, February 22, 2008, http://www.doi.gov/news/08_News_Releases/080222_fact_sheet.html.
45. *Shivwits Band of Paiute Indians v. Utah*, 428 F.3d 966 (2005); and "State Cannot Regulate Billboards on Indian Trust Land," *Land Use Law Report* 33 (November 16, 2005): 171–72.

Chapter Seven

1. The book was Mark Monmonier, *Bushmanders and Bullwinkles: How Politicians Manipulate Electronic Maps and Census Data to Win Elections* (Chicago: University of Chicago Press, 2001), and the review was Richard L. Berke, "Don't Know Much Geography: An Examination of How Creative Cartography Affects U.S. Politics," *New York Times*, book review section, May 27, 2001.
2. James T. Austin, *The Life of Elbridge Gerry, with Contemporary Letters to the Close of the American Revolution* (Boston: Wells and Lilly, 1829; reprint, New York: Da Capo Press, 1970), 2:347.
3. John Ward Dean, "The Gerrymander," *New England Historical and Genealogical Register* 46 (1892): 374–83.
4. Monmonier, *Bushmanders and Bullwinkles*, esp. 2–8 and 35–63.
5. Ibid., 64–76 and 104–20.
6. Douglas J. Amy, *Real Choices/New Voices: How Proportional Representation Elections Could Revitalize American Democracy*, 2nd ed. (New York: Columbia University Press, 2002).

7. For concise descriptions of the process, see Howard Fain, "P.R. Elections in Cambridge, Mass.," *National Civic Review* 83 (1994): 84–85; and Monmonier, *Bushmanders and Bullwinkles*, 138–39.

8. (3,000,000 ÷ (3 + 1)) + 1 = (3,000,000 ÷ 4) + 1 = 750,000 + 1 = 750,001.

9. Cambridge further simplifies the process by eliminating any candidate with fewer than 50 votes.

10. According to the Cambridge Elections Commission Web site, "the same process formerly carried out manually by more than a hundred counters over the course of a week is conducted in a matter of seconds by the electronic sorting, counting, and transfer of votes." As elsewhere, additional time is needed to accommodate write-in votes and ballots "marked in a way that could not be read by the scanner at the precinct level." City of Cambridge Elections Commission, Computerized Tabulation Process, http://www.cambridgema.gov/election/computerized_tabulation.cfm (accessed April 20, 2008).

11. See, for example, California Reform Initiative, Reforming Redistricting through Superdistricts, http://www.fairvote.org/ca/?page=839 (accessed April 23, 2008).

12. See FairVote Program for Representative Government, North Carolina, http://www.fairvote.org/media/research/superdistricts/NorthCarolina.pdf (accessed April 21, 2008).

13. See FairVote Program for Representative Government, Pennsylvania, http://www.fairvote.org/media/research/superdistricts/Pennsylvania.pdf (accessed April 21, 2008).

14. For a summary of objections not related to computation, see Amy, *Real Choices/New Voices*, 186–214.

15. Oddly, the provision requiring single-member districts was an amendment to a private immigration bill, passed as http://www.fairvote.org/library/history/flores/district.htm-_ftnref175*For the Relief of Dr. Ricardo Vallejo Samala*, HR 2275, 90th Cong., 1st sess. See "House District Standards," *Congressional Quarterly Weekly Report* 25 (1967): 2475. Ostensibly a civil rights measure, the ban was intended to prevent winner-take-all at-large elections, which could dilute the clout of newly enfranchised African Americans in the South. See Tory Mast, "The History of Single Member Districts for Congress," FairVote Program for Representative Government, http://www.fairvote.org/?page=526 (accessed April 22, 2008). There are no constitutional obstacles to multimember districts per se. See Richard K. Scher, Jon L. Mills, and John J. Hotaling, *Voting Rights and Democracy: The Law and Politics of Districting* (Chicago: Nelson-Hall, 1997), 293–94.

16. A positive sign is a 2006 referendum in which Minneapolis voters adopted "instant-runoff voting," similar to choice voting, to elect their mayor and city council. Terry Collins, "Much Work Ahead for Instant-Runoff Voting," *Minneapolis Star Tribune*, November 15, 2006.

Chapter Eight

1. Amy E. Hillier, "Spatial Analysis of Historical Redlining: A Methodological Exploration," *Journal of Housing Research* 14 (2003): 137–67, esp. 139.
2. "Excerpts from Johnson's Message to Congress on Housing and Urban Problems," *New York Times*, February 23, 1968.
3. Charles Abrams, "Slums, Ghettos, and the G.O.P.'s Remedy," *The Reporter* 10 (May 11, 1954): 27–30; quotation on 29. Abrams did not mention the HOLC by name but condemned its appraisal methods, as adopted by the FHA.
4. Rosalind Tough, "The Life Cycle of the Home Owners' Loan Corporation," *Land Economics* 27 (1951): 324–31.
5. Kristen B. Crossney and David W. Bartelt, "Residential Security, Risk, and Race: The Home Owner's Loan Corporation and Mortgage Access in Two Cities," *Urban Geography* 26 (2005): 707–36.
6. Amy Elizabeth Hillier, "Redlining and the Home Owners' Loan Corporation" (PhD diss., University of Pennsylvania, 2001), 84–97.
7. Ibid., 118–19, 170.
8. For Jackson's discovery, see Hillier, "Redlining and the Home Owners' Loan Corporation," 8, 34n1. Also see Kenneth T. Jackson, *Crabgrass Frontier: The Suburbanization of the United States* (New York: Oxford University Press, 1985); quotation on 197.
9. Douglas S. Massey and Nancy A. Denton, *American Apartheid: Segregation and the Making of the Underclass* (Cambridge, MA: Harvard University Press, 1993), 52.
10. Buzz Bissinger, *A Prayer for the City* (New York: Random House, 1997), 206
11. Douglas W. Rae, *City: Urbanism and Its End* (New Haven, CT: Yale University Press, 2003), 265.
12. I wanted to include an example, but experimentation with the Syracuse map confirmed that red tints don't reproduce well in black and white. Even so, it's easy to imagine the maps' appearance and the plausible impact.
13. Crossney and Bartelt, "Residential Security, Risk, and Race," 735.
14. Ibid., 736.
15. John T. Metzger, "Planned Abandonment: The Neighborhood Life-Cycle Theory and National Urban Policy," *Housing Policy Debate* 11 (2000): 7–40.
16. Homer Hoyt, *The Structure and Growth of Residential Neighborhoods in American Cities* (Washington, DC: Federal Housing Administration, 1939), 77. Hoyt's diagram portrays generalized patterns for 30 cities.
17. Ibid., 114.
18. Ibid., 47.
19. Ibid., 48.
20. Ibid., 132.
21. Amy E. Hillier, "Redlining and the Home Owners' Loan Corporation," *Journal of Urban History* 29 (2003): 394–420, esp. 397.

22. Ibid., 398–400.

23. Ibid., 414.

24. The acronym stands for Zone Improvement Program. For a discussion of postal-code characterization of neighborhoods, see Mark Monmonier, *Spying with Maps: Surveillance Technology and the Future of Privacy* (Chicago: University of Chicago Press, 2002), 140–53.

25. Editorial, *Syracuse Post-Standard*, March 7, 1997.

26. Nina Kim, "Car Rental Business Won't Get Permits," *Syracuse Post-Standard*, April 2, 1998.

27. Maureen Sieh, "Rental Car Co. Settles Bias Suit," *Syracuse Post-Standard*, July 30, 1999.

28. Kenneth R. Harney, "'Declining Market' Labeling under Fire," *Syracuse Post-Standard*, real estate section, April 27, 2008.

29. J. Robert Hunter, State Automobile Insurance Regulation: A National Quality Assessment and In-Depth Review of California's Uniquely Effective Regulatory System (Washington, DC: Consumer Federation of America, 2008), 38. Online at www.consumerfed.org/pdfs/state_auto_insurance_report .pdf (accessed May 14, 2008).

30. The concept of enterprise zones was developed in Britain in the mid-1970s, and imported to the United States in the early 1980s. For an overview of the concept, see Stuart M. Butler, *Enterprise Zones: Greenlining the Inner Cites* (New York: Universe Books, 1981). For a postmortem on Britain's experience—the program was phased out in 2006—see Colin Jones, "Verdict on the British Enterprise Zone Experiment," *International Planning Studies* 11 (2006): 109–23.

31. Alan H. Peters and Peter S. Fisher, *State Enterprise Zone Programs: Have They Worked?* (Kalamazoo, MI: Upjohn Institute for Economic Research, 2002), 2.

32. U.S. Government Accountability Office, Empowerment Zone and Enterprise Community Program: Improvements Occurred in Communities, but the Effect of the Program Is Unclear, Report GAO-06-727, September 2006; quotation on 29.

33. Robert T. Greenbaum, "Siting It Right: Do States Target Economic Distress When Designating Enterprise Zones?" *Economic Development Quarterly* 18 (2004): 67–80; quotation on 79.

34. Barclay Gibbs Jones and Donald M. Manson, "The Geography of Enterprise Zones: A Critical Analysis," *Economic Geography* 58 (1982): 329–42.

35. Emily N. Spiegel, "Empire Zones: Cultivating New York Businesses," *CPA Journal* 72 (January 2002): 22–27. For status and a description of requirements, see Empire Zones, http://www.nylovesbiz.com/Tax_and_Financial_ Incentives/Empire_Zones/default.asp (accessed May 17, 2008).

36. For a devastating critique, see Citizens Budget Commission [New York], "It's Time to End New York's Empire Zone Program," December 2008, http:// www.cbcny.org/Ending_Empire_Zones.pdf (accessed January 2, 2009).

37. This example is hardly hypothetical: companies with "fewer than two employees" claimed over $46 million in tax reduction credits in 2005. Mike McAndrew, "Anatomy of Yet Another Loophole," *Syracuse Post-Standard*, June 3, 2007.

38. Mike McAndrew and Michelle Breidenbach, "Seven Years of Abuse Stops Here," *Syracuse Post-Standard*, February 17, 2008.

39. Mike McAndrew and Michelle Breidenbach, "Tax Breaks Revealed for Empire Zone Companies," *Syracuse Post-Standard*, March 14, 2007.

40. For examples, see Mike McAndrew, "Cortland Wants to Kick Firm Out of Empire Zone," *Syracuse Post-Standard*, October 7, 2007; and Aaron Gifford, "Madison County Firm Loses Empire Zone Status," *Syracuse Post-Standard*, December 17, 2007.

41. Quoted in Delen Goldberg, "Groups Testify to Empire Zone Failure," *Syracuse Post-Standard*, December 19, 2007.

42. New York State Commissioner of Economic Development, Revised Regulations, Parts 10–14, http://www.nylovesbiz.com/pdf/EmpireZonesRegulations.pdf (accessed May 19, 2008).

43. Ibid, 10.4(3)(i), p. 16.

44. See the Empire Zone map online at City of Geneva Department of Planning and Economic Development, http://www.genevadevelopment.gov office2.com/ (accessed May 21, 2008).

45. Michelle Breidenbach, "Communities Sell Empire Zone Rights to Companies for Millions," *Syracuse Post-Standard*, November 19, 2006.

46. Ibid. The village Web site includes a map that shows Empire Zone territory within its border but is silent about outside allocations. See Potsdam Office of Planning and Development, http://www.vi.potsdam.ny.us/econ/EZmap 3-07-07.pdf (accessed May 21, 2008).

47. New York State Commissioner of Economic Development, Revised Regulations, 10.8(e), p. 27.

48. For the list of Empire Zone businesses that claimed tax breaks in 2005, see blog.syracuse.com/news/EZTaxCreditAmounts2005.xls (accessed May 19, 2008); for the list of companies that missed goals, see blog.syracuse.com/ newstracker/2007/07/ESD list for EZ letters.xls (accessed May 19, 2007). Also see Michelle Breidenbach, "3,000 Warning Letters Sent to Empire Zone Users," *Syracuse Post-Standard*, July 31, 2007.

49. Greater Syracuse Economic Growth Council, Hamlet of Jamesville Inset Map, http://www.syracusecentral.com/empire_zone/pdf/jamesville.pdf (accessed May 21, 2008).

Chapter Nine

1. John O'Brien, "Great Scott House Finally Buys the Farm," *Syracuse Post-Standard*, December 4, 1991; and John O'Brien, "Zoning Rebel Now Wants to Be Enforcer," *Syracuse Post-Standard*, May 12, 2008.
2. George B. Ford, "City Planning by Coercion or Legislation," *American City* 14 (1916): 328–33; quotation on 333.
3. For examples, see Herbert S. Swan, "Does Your City Keep Its Gas Range in the Parlor and Its Piano in the Kitchen?" *American City* 22 (1920): 339–44.
4. *Village of Euclid v. Ambler Realty Co.*, 272 U.S. 365 at 397 (1926).
5. Donald L. Elliott, *A Better Way to Zone: Ten Principles to Create More Livable Cities* (Washington, DC: Island Press, 2008), 91.
6. New York City Department of City Planning, About NYC Zoning, http://www.nyc.gov/html/dcp/html/zone/zonehis.shtml (accessed June 14, 2008). The city Web site reports 42 stories; other sources say 40, 39, and 36. On an image of the building I counted no more than 36 lines of windows above ground.
7. Keith D. Revell, *Building Gotham: Civic Culture and Public Policy in New York City, 1898–1938* (Baltimore: John Hopkins University Press, 2003), 202–4.
8. For a concise summary of the city's special-purpose districts and their unique needs, see TenentNet's *NYC Zoning Handbook, Chapter 11: Special Zoning Districts*, http://www.tenant.net/Other_Laws/zoning/zonch11.html (accessed June 17, 2008).
9. New York Department of City Planning, *Zoning Resolution, Article VIII: Special Purpose Districts, Chapter 1: Special Midtown District* (New York, 2008); quotations on 81-00 and 81-01.
10. New York Department of City Planning, Environmental Review (E) Designations, http://www.nyc.gov/html/dcp/html/env_review/e_designation_faqs.shtml (accessed June 18, 2008).
11. New York Department of City Planning, About Zoning: How Zoning Is Administered and Amended, http://home2.nyc.gov/html/dcp/html/zone/zonehis.shtml (accessed June 18, 2008).
12. Roger Bernhardt and Ann M. Burkhart, *Real Property in a Nutshell*, 5th ed. (St. Paul, MN: Thomson West, 2005), 399.
13. Tim Krause, "DeWitt Eyed for Huge Energy Plant," *Syracuse Post-Standard*, December 17, 2006.
14. Frederic Pierce, "DeWitt Board Will Oppose Coal Plant," *Syracuse Post-Standard*, May 23, 2007.
15. Jamesville Zoning Proposals, Town of DeWitt Zoning Amendment, draft May 15, 2008; online at www.townofdewitt.com (assessed June 17, 2008).
16. Jamesville Zoning Committee, "Jamesville Hamlet Information Hearing June 17, 2008," three-page handout posted on the town Web site, www.townofdewitt.com (accessed June 16, 2008).

17. Tami Zimmerman, "DeWitt: New Zones Approved," *DeWitt Times*, 18 June 2008.

18. *Code of Federal Regulations,* title 36, part 60, section 3(d) (July 1, 2007 edition). In addition, "a district may also comprise individual elements separated geographically but linked by association or history."

19. For examples, see Karen Eschbacher, "In the Grasp of History on Cape, Building Restrictions Draw Residents' Ire," *Boston Globe,* June 4, 2000.

20. For insights on the diverse sharpness of historic district boundaries, see John P. Conron et al., *Delineating Edges of Historic Districts* (Washington, DC: Preservation Press, National Trust for Historic Preservation, 1976).

21. Donna J. Seifert, *Defining Boundaries for National Register Properties,* National Register Bulletin 21 (Washington, DC: U.S. Dept. of the Interior, National Park Service, Interagency Resources Division, National Register of Historic Places, 1995), 2.

22. Elizabeth A. Lunday, "Who's in Control Here?" *Planning* 71 (June 2005): 18–23.

23. For insights on the diverse perceptions of historic districts and their effects, see Tad Heuer, "Living History: How Homeowners in a New Local Historic District Negotiate Their Legal Obligations," *Yale Law Journal* 116 (2007): 768–822.

24. Robin M. Leichenko, N. Edward Coulson, and David Listokin, "Historic Preservation and Residential Property Values: An Analysis of Texas Cities," *Urban Studies* 38 (2001): 1973–87.

25. Nicholas Blomley, *Unsettling the City: Urban Land and the Politics of Property* (New York: Routledge, 2004), 81–82.

26. Prudent municipalities zone the floodway and its fringe for less intensive use. James M. Holway and Raymond J. Burby, "Private Markets, Public Decisions: An Assessment of Local Land-Use Controls for the 1990s," *Land Economics* 66 (1990): 259–71.

27. For a discussion of approaches, see Mustafa Yildirim and Bülent Topkaya, "Groundwater Protection: A Comparative Study of Four Vulnerability Mapping Methods," *CLEAN—Soil, Air, Water* 35 (2007): 594–600.

28. National Research Council, Committee on Characterization of Wetlands, *Wetlands: Characteristics and Boundaries* (Washington, DC: National Academy Press, 1995), 43.

29. For a sense of the complexity of wetland delineation, see Ralph W. Tiner, *Wetland Indicators: A Guide to Wetland Identification, Delineation, Classification, and Mapping* (Boca Raton, FL: Lewis Publishers, 1999).

30. Ralph W. Tiner, *In Search of Swampland: A Wetland Sourcebook and Field Guide* (New Brunswick, NJ: Rutgers University Press, 1998), 6.

31. The Fish and Wildlife Service transferred responsibility for the list to the Army Corps of Engineers in late 2006 but still provides access through its Web site. See U.S. Fish and Wildlife Service, Wetland Plants, http://www.fws.gov/nwi/Plants/plants.htm (accessed June 25, 2008).

32. Margaret N. Strand and Lowell Rothschild, "Wetlands: Avoiding the Swamp Monster," in *Environmental Aspects of Real Estate and Commercial Transactions: From Brownfields to Green Buildings,* ed. James B. Witkin, 737–79 (Chicago: American Bar Association, 2004).

33. James V. DeLong, *Property Matters: How Property Rights Are under Assault—and Why You Should Care* (New York: Free Press, 1997), 134.

34. *Rapanos v. United States,* 547 U.S. 715 (2006). Also see Linda Greenhouse, "Justices Divided on Protections over Wetlands," *New York Times,* June 20, 2006; and Donald Kennedy and Brooks Hanson, "What's a Wetland, Anyhow?" *Science* 313 (2006): 1019.

35. John M. Broder, "After Lobbying, Wetlands Rules Are Narrowed," *New York Times,* July 6, 2007. Also see Samuel P. Bickett, "The Illusion of Substance: Why Rapanos v. United States and Its Resulting Regulatory Guidance Do Not Significantly Limit Federal Regulation of Wetlands," *North Carolina Law Review* 86 (2008): 1032–46.

36. Associated Press, "Ga. Man Sentenced for Faking Maps," *Durham (NC) Herald-Sun,* February 26, 2006; US EPA Region 4: Environmental Accountability, Enforcement Actions Archives 2006 and 2005, CWA 2006 (Clean Water Act), http://www.epa.gov/Region4/ead/general/recentarchives.html (accessed June 24, 2008); and Enviro.BLR.com, "Surveyor Pleads Guilty in North Carolina Wetlands Case," November 30, 2005, http://enviro.blr.com/display.cfm/id/62604 (accessed June 24, 2008).

Chapter Ten

1. *Schad v. Borough of Mount Ephraim,* 452 U.S. 61 (1981).

2. Linda Greenhouse, "High Court Rejects Ban on Live Entertainment," *New York Times,* June 2, 1981.

3. *Schad v. Borough of Mount Ephraim,* 452 U.S. 61 at 79 (1981).

4. Eric Damian Kelly and Connie Cooper, *Everything You Always Wanted to Know about Regulating Sex Businesses* (Chicago: American Planning Association, 2000), 67.

5. Ibid., 65.

6. "Ordinance Left No Location in City for Adult Business," *Land Use Law Report* 34, no. 12 (June 28, 2006): 96.

7. See §192—66, B(2) in Zoning: Chapter 192 of the Town Code, DeWitt, New York, available through www.townofdewitt.com (assessed July 9, 2008).

8. Kelly and Cooper, *Everything You Always Wanted to Know about Regulating Sex Businesses,* 117.

9. Ibid., 35.

10. John Doherty, "Officials Consider Oxbow Road for an Adult Zone," *Syracuse Post-Standard,* December 7, 2006; and Ngoc Huynh, "Cicero Selects Adult Zone," *Syracuse Post-Standard,* December 29, 2006.

11. Jim Read, "Manlius Residents Concerned about Adult Zone," *Syracuse Post-Standard,* January 11, 2007.

12. Town of Cicero, New York, municipal code, Zoning: Chapter 210, February 2008 printing, 39–51.

13. Ibid., 50.

14. Jules B. Gerard, *Local Regulation of Adult Businesses* (St. Paul, MN: Thomson/West, 2004); 241–50, quotation on 245.

15. Read, "Manlius Residents Concerned about Adult Zone."

16. Kelly and Cooper, *Everything You Always Wanted to Know about Regulating Sex Businesses,* 86.

17. "Cleanup Comes to the Combat Zone," *New York Times,* July 20, 1989.

18. Mara Vorhees and John Spelman, *Boston: City Guide,* 3rd ed. (Oakland, CA: Lonely Planet, 2007), 95.

19. *Central Avenue, Inc. v. The City of Charlotte,* 2006 U.S. Dist. Lexis 68074 at 68090 (W.D.N.C.). Also see "Sexually Oriented Businesses Contend Zoning Ordinances Put First Amendment Rights in a Compromising Position," *Land Use Law Report* 34, no. 16 (October 2006): 135–36.

20. New York State Division of Criminal Justice Services, Sex Offender Registry, http://www.criminaljustice.state.ny.us/nsor/index.htm (accessed July 14, 2008).

21. Massachusetts Sex Offender Registry Board, http://www.mass.gov/?pageID =eopstopic&L=3&L0=Home&L1=Crime+Prevention+%26+Personal +Safety&L2=Sex+Offenders&sid=Eeops (accessed July 14, 2008).

22. Georgia Bureau of Investigation, Georgia Sex Offender Registry, http:// gbi.georgia.gov/00/channel_modifieddate/0,2096,67862954_87983024,00 .html (accessed July 14, 2008).

23. Office of the Attorney General of Ohio, Electronic Sex Offender Notification and Registration, http://www.esorn.ag.state.oh.us/Secured/p21_2.aspx.

24. Margaret Trioa, "Ohio's Sex Offender Residency Restriction Law: Does It Protect the Health and Safety of the State's Children or Falsely Make People Believe So?" *Journal of Law and Health* 19 (2004/2005): 331–70.

25. Council of State Governments, Sex Offender Residency Restriction Laws: 2008, http://www.csg.org/policy/pubsafety/documents/SOResRestriction laws2008.pdf (accessed July 15, 2008).

26. Joseph L. Lester, "Off to Elba! The Legitimacy of Sex Offender Residence and Employment Restrictions," *Akron Law Review* 40 (2007): 339–89; and Richard Tewksbury, "Exile at Home: The Unintended Collateral Consequences of Sex Offender Residency Restrictions," *Harvard Civil Rights– Civil Liberties Law Review* 42 (2008): 531–40.

27. Laura Mansnerus, "Zoning Laws That Bar Pedophiles Raise Concerns for Law Enforcers," *New York Times,* November 27, 2006.

28. John Doherty, "Cicero Restricts Sex Offenders," *Syracuse Post-Standard,* November 30, 2006.

29. John Doherty, "Officials Spar over Sex-Offender Law," *Syracuse Post-Standard*, December 19, 2006.

30. Minnesota Department of Corrections, "Level Three Sex Offenders Residential Placement Issues: 2003 Report to the Legislature," January 2003, revised February 2004; quotations on 9, 10.

31. Jake Palmateer "Otsego to Review Waivers for Sex Offenders," *Oneonta (N.Y.) Daily Star*, December 12, 2007. Quotation from Otsego County, New York, Board of Representatives, County Board Minutes, July 5, 2007, http://www.otsegocounty.com/depts/bor/documents/Minute07-05-2007.pdf (accessed July 16, 2008).

32. The Hat City Blog chronicled the enactment and contentious implementation of Danbury's "Child Safety Zone" law. For an example of the signs, which not all sites received, see "Sex Offender Flashback: What's Wrong with this Picture?" July 1, 2008, http://hatcityblog.blogspot.com/2008/07/sex-offender-saga-whats-wrong-with-this.html. For discussion of "zone hopping" and the effort to remover the loophole, see "The Sex Offender Saga: 1 Year, 6 Months, 26 Days," July 1, 2008, http://hatcityblog.blogspot.com/2008/07/sex-offender-saga-1-year-6-months-26.html. (both accessed July 16, 2008).

33. Marisa L. Mortensen, "GPS Monitoring: An Ingenious Solution to the Threat Pedophiles Pose to California's Children," *Journal of Juvenile Law* 27 (2006): 17–32; and R. Brooks Whitehead, "Good for More than Just Driving Directions: GPS Helps Protect Californians from Recidivist Sex Offenders," *McGeorge Law Review* 38 (2007): 265–75.

34. Pamela Foohey, "Applying the Lessons of GPS Monitoring of Batterers to Sex Offenders," *Harvard Civil Rights–Civil Liberties Law Review* 43 (2008): 281–84.

Chapter Eleven

1. See, for example, Ken Belson, "Old Sewer Mapping System Undergoes a Welcome Update," *New York Times*, August 19, 2008.

2. U.S. Department of Transportation, Office of Pipeline Safety, *Common Ground: Study of One-Call Systems and Damage Prevention Best Practices* (Washington, DC, 1999).

3. Hyung Seok Jeong and Duley M. Abraham, "A Decision Tool for the Selection of Imaging Technologies to Detect Underground Infrastructure," *Tunnelling and Underground Space Technology* 19 (2004): 175–91.

4. Dana Bell, *Smithsonian Atlas of World Aviation* (New York: Collins, 2008), 198–215.

5. Prohibited areas and restricted areas are considered Special Use Airspace. For a Special Use Airspace listing, see the FAA Air Traffic Plans and Pub-

lications Web site at http://www.faa.gov/airports_airtraffic/air_traffic/ publications. Information for this and the following paragraph are based on publication JO FFA Order 7400.8P, dated February 16, 2008.

6. Helene Cooper, "This President's Escape Is Sweet Home Chicago," *New York Times,* February 16, 2009.

7. U.S. War Department, *Conventional Signs: United States Army Maps, with Supplement Containing Changes Nos. 1 and 2* (Washington, DC: Government Printing Office, 1918); quotation on iii. Aeronautical symbols were inserted as "Changes No. 2," effective December 3, 1918. The original edition, issued in 1912, made no mention of aeronautical charts.

8. Ralph E. Ehrenberg, "Up in the Air in More Ways than One: The Emergence of Aeronautical Charts in the United States," in *Cartographies of Travel and Navigation,* ed. James R. Akerman, 207–59 (Chicago: University of Chicago Press, 2006).

9. "Report on Air Navigation Charts Adopted," *Air Commerce Bulletin* 1, no. 11 (December 2, 1929): 27–28.

10. Board of Surveys and Maps of the Federal Government, "Report on Standard Symbols for Air Navigation Maps by the Committees on Technical Standards and Aerial Navigation Maps," September 28, 1928. Also see Raymond L. Ross, "Mapping United States Airways," *Military Engineer* 20 (1928): 476–78.

11. "Board on Air-Space Reservations Constituted," *Air Commerce Bulletin* 1, no. 18 (March 15, 1930): 7–8.

12. "Flying Rules for Certified High-Explosive Danger Areas and Instructions for Marking Obstructions to Air Navigation," *Air Commerce Bulletin* 1, no. 22 (May 15, 1930): 8–10.

13. "Designation of Air Space over District of Columbia as Prohibited Area on March 4, 1933." *Air Commerce Bulletin* 4, no. 17 (March 1, 1933): 428.

14. "Special Air Traffic Rule Prohibits Flight over Downtown Washington, D.C." *Air Commerce Bulletin* 6, no. 10 (April 15, 1935): 238.

15. "Map of Prohibitive Area over Washington, D.C.," *Air Commerce Bulletin* 6, no. 11 (May 15, 1935): 268.

16. Federal Aviation Administration, Special Use Airspace and Air Traffic Control Assigned Airspace Web site, http://sua.faa.gov/sua/.

17. For a crude map of the ADIZs bracketing the conterminous United States, see Federal Aviation Administration, *Aeronautical Information Manual: Official Guide to Basic Flight Information and ATC Procedures,* February 16, 2006 ed. (Washington, DC: U.S. Department of Transportation, 2006), page 5-6-7. ADIZs are also defined for Alaska, Hawaii, and Guam.

18. Michele S. Sheets, "From ADIZ to SFRA: The FAA's Compliance with Administrative Procedures to Codify Washington, DC Flight Restrictions," *Journal of Air Law and Commerce* 71 (2006): 615–49.

19. Nicholas J. Wheeler, *Saving Strangers: Humanitarian Intervention in International Society* (New York: Oxford University Press, 2000), esp. 161–68.

20. Elaine Sciolino, "U.S. Warns against Attack by Iraq on Kurdish Refugees," *New York Times*, April 11, 1991.

21. Michael R. Gordon, "British, French, and U.S. Agree to Hit Iraqi Aircraft in the South," *New York Times*, August 19, 1992.

22. Timothy P. McIlmail, "No-Fly Zones: The Imposition and Enforcement of Air Exclusion Regimes over Bosnia and Iraq," *Loyola of Los Angeles International and Comparative Law Review* 17 (1994): 35–83.

23. Arnold Beichman, *The Long Pretense: Soviet Treaty Diplomacy from Lenin to Gorbachev* (New Brunswick, NJ: Transaction Publishers, 1991), 157–58.

24. Ibid., 158.

25. Sue Davis, *The Russian Far East: The Last Frontier?* (London: Routledge, 2003), 105.

26. "Travel Regulations for Soviet Citizens in the United States" [press release dated January 3], *Department of State Bulletin* 32 (1955): 193–97.

27. "U.S. Revises Areas Closed to Soviet Citizens; proposes Mutual Abolition or Reduction of Restrictions" [press release dated November 12], *Department of State Bulletin* 49 (1963): 855–59. My counts for 1963 and 1983 include independent cities in Virginia and other areas listed along with counties.

28. Bernard Gwertzman, "State Dept. Alters the Rules for Russians' Travel in U.S.," *New York Times*, November 20, 1983.

29. "Executive Order Authorizing the Secretary of War to Prescribe Military Areas," *Federal Register* 7, no. 33 (February 25, 1942): 1407.

30. Commission on Wartime Relocation and Internment of Civilians, *Personal Justice Denied: Report of the Commission on Wartime Relocation and Internment of Civilians* (Washington, DC: U.S. Government Printing Office, 1983), 72.

31. New York State Department of Environmental Conservation, Wildlife Management Units, http://www.dec.ny.gov/outdoor/8302.html (accessed November 5, 2008).

32. New York State Department of Environmental Conservation, Deer Management Permits, http://www.dec.ny.gov/outdoor/6403.html (accessed November 5, 2008).

33. California Department of Food and Agriculture, "Oriental Fruit Fly Quarantine Declared in Lakewood Area," Press Release no. 08-053, August 26, 2008.

Chapter Twelve

1. For an overview see Anthony LaMarca and Eyal de Lara, *Location Systems: An Introduction to the Technology behind Location Awareness* (San Rafael, CA: Morgan & Claypool, 2008); and Mark Monmonier, "Keeping Track," in his *Spying with Maps: Surveillance Technologies and the Future of Privacy* (Chicago: University of Chicago Press, 2002), 125–39.

2. The automatically administered soporific seems a more reasonable response

when a subject tries to remove the tracking device; see Max Winkler, "Walking Prisons: The Developing Technology of Electronic Controls," *The Futurist* 27 (July/August 1993): 34–36.

3. For discussion of pre-electronic "protest exclusion zones," see Lynn A. Staeheli and Don Mitchell, *The People's Property?: Power, Politics, and the Public* (New York: Routledge, 2008), esp. 15–19.

4. Jerome E. Dobson, "What Are the Ethical Limits of GIS?" *GeoWorld* 13 (May 2000): 24–25.

5. Jerome E. Dobson and Peter F. Fisher, "Geoslavery," *IEEE Technology & Society Magazine* 22 (Spring 2003): 47–52; quotation on 47–48.

6. See, for example, Kathy Chu, "Drivers Exchange their Privacy for Auto Insurance Savings," *Globe and Mail*, August 19, 2004; and Kevin Maney, "For a Price, Would You Let Car Insurer Tag Along for the Ride?" *USA Today*, August 3, 2005.

7. For examples, see Liam Ford, "GPS May Help Enforce Restraining Orders," *Chicago Tribune*, April 2, 2008.

8. Mark Monmonier, "Geolocation and Locational Privacy: The 'Inside' Story on Geospatial Tracking," in *Privacy and Technologies of Identity: A Cross-Disciplinary Conversation*, ed. Katherine J. Strandberg and Daniela Stan Raicu, 75–91 (New York: Springer Science + Business Media, 2006).

9. Don Herskovitz, "GPS Insurance: Antijamming the System," *Journal of Electronic Defense* 23 (December 2000): 41–45.

10. David Hughes, "New Age Anti-Collision," *Aviation Week and Space Technology* 169 (July 14, 2008): 163–64; and J. R. Wilson, "The Time Is Now for Sense and Avoid," *Aerospace America* 44 (June 2006): 32–36.

11. David Glenn Bird and James M. Janky, On-board Apparatus for Avoiding Restricted Air Space in Non-overriding Mode, U.S. Patent 6,675,095, filed December 15, 2001, and issued January 6, 2004.

12. Michael Zuschlag, *Potential Interventions by Government and Industry to Minimize Violations of Temporary Flight Restrictions*, report NASA/CR—2005-213924, NASA Scientific and Technical Information (STI) Program Office (September 2005); quotation on 18.

13. "Positive Signals," *Engineer* 293 (February 2005): 8; and Don Phillips, "Digital Railroad," *Technology Review* 105 (March 2002): 74–79.

14. For examples, see John R. Quinn, "Warning! Traffic Jam Straight Ahead," *New York Times*, October 30, 2008, Cars section.

15. For discussion of counter mapping, see Jay T. Johnson, Renee Pualani Louis, and Albertus Hadi Pramono, "Facing the Future: Encouraging Critical Cartographic Literacies in Indigenous Communities," *ACME: An International E-Journal for Critical Geographies* 4 (2006): 80–98; and John Pickles, "Counter-Mappings: Cartographic Reason in the Age of Intelligent Machines and Smart Bombs," in his *A History of Spaces: Cartographic Reason, Mapping and the Geo-Coded World* (London: Routledge, 2004), 179–94.

16. John C. Baker and others, *Mapping the Risks: Assessing the Homeland Security Implications of Publicly Available Geospatial Information* (Santa Monica, CA: RAND Corporation, 2004).

17. Vicki Haddock, "A Whole New Way to Look at the World: Satellite Imagery Turns Globe into a Computer Peep Show," *San Francisco Chronicle*, April 1, 2007.

SELECTED
READINGS FOR
FURTHER
EXPLORATION

Readers seeking additional information about the impact of restrictive maps as well as the mapmakers and bureaucrats engaged in prohibitive cartography might appreciate this diverse list of references, organized by chapter.

Keep Off!

Conover, Milton. *The General Land Office: Its History, Activities and Organization.* Baltimore: Johns Hopkins Press, 1923.

Hubbard, Bill, Jr. *American Boundaries: The Nation, the States, the Rectangular Survey.* Chicago: University of Chicago Press, 2009.

Johnson, Hildegard Binder. *Order upon the Land: The U.S. Rectangular Land Survey and the Upper Mississippi Country.* New York: Oxford University Press, 1976.

Kain, Roger J. P. "The Role of Cadastral Surveys and Maps in Land Settlement from England." *Landscape Research* 27 (2002): 11–24.

Kain, Roger J. P., and Elizabeth Baigent, eds. *The Cadastral Map in the Service of the State: A History of Property Mapping.* Chicago: University of Chicago Press, 1992.

Kaplan, Bernard M. *A Comprehensive Guide to Modern Real Estate Law Practice and Brokerage.* Chicago: Commerce Clearing House, 1989.

Marschner, F. J. *Land Use and Its Patterns in the United States.* U.S. Department of Agriculture Handbook no. 153. Washington, DC: Government Printing Office, 1958.

National Research Council, Committee on Land Parcel Databases: A National Vision. *National Land Parcel Data: A Vision for the Future.* Washington, DC: National Academies Press, 2007.

Robillard, Walter G., Donald A. Wilson, and Curtis M. Brown. *Brown's Boundary Control and Legal Principles.* 5th ed. New York: John Wiley and Sons, 2003.

Scott, James C. *Seeing Like a State: How Certain Schemes to Improve the Human Condition Have Failed.* New Haven, CT: Yale University Press, 1998, esp. 36–58.

U.S. Bureau of Land Management. *Manual of Instructions for the Survey of the Public Lands of the United States.* 1973 ed. Technical Bulletin no. 6. Washington, DC: Government Printing Office, 1973.

Wattles, Gurdon H. *Writing Legal Descriptions in Conjunction with Survey Boundary Control.* Orange, CA: G. H. Wattles Publications, 1979.

Keep Out!

Anderson, Ewan W. *International Boundaries: A Geopolitical Atlas.* New York: Routledge, 2003.

Blake, Gerald. "Boundary Permeability in Perspective." In *Holding the Line: Borders in a Global World,* ed. Heather N. Nicol and Ian Townsend-Gault, 15–25. Vancouver: UBC Press, 2005.

Cohen, Shaul E. "Israel's West Bank Barrier: An Impediment to Peace?" *Geographical Review* 96 (2006): 682–95.

Edwards, Thomas M. "Information Geopolitics: Blurring the Lines of Sovereignty." In *Holding the Line: Borders in a Global World,* ed. Heather N. Nicol and Ian Townsend-Gault, 26–49. Vancouver: UBC Press, 2005.

Hendrix, Burke A. "Moral Minimalism in American Indian Land Claims." *American Indian Quarterly* 29 (2005): 538–59.

Holdich, Thomas H. *Political Frontiers and Boundary Making.* London: Macmillan and Co., 1916.

Jones, Stephen B. "The Description of International Boundaries." *Annals of the Association of American Geographers* 33 (1943): 99–117.

Newman, David. "Boundary Geopolitics: Toward a Theory of Territorial Lines?" In *Routing Borders between Territories, Discourses and Practices,* ed. Eiki Berg and Henk van Houtum, 277–91. Burlington, VT: Ashgate, 2003.

———. "The Lines That Continue to Separate Us: Borders in Our 'Borderless' World." *Progress in Human Geography* 30 (2006): 143–61.

Wilson, Larman C. "The Settlement of Boundary Disputes: Mexico, the United

States, and the International Boundary Commission." *International and Comparative Law Quarterly* 29 (1980): 38–53.

Absentee Landlords

Collier, Peter. "Imperial Boundary Making in the 19th Century." *Society of Cartographers Bulletin* 38, no. 2 (2004): 45–47.

Crampton, Jeremy W. "The Cartographic Calculation of Space: Race Mapping and the Balkans at the Paris Peace Conference of 1919." *Social and Cultural Geography* 7 (2006): 731–52.

Dodds, Klaus. *Geopolitics in Antarctica: Views from the Southern Ocean Rim.* Chichester, UK: John Wiley and Sons, 1997.

Dzurek, Daniel J. "What Makes Territory Important: Tangible and Intangible Dimensions." *Geoforum* 64 (2005): 263–74.

Joyner, Christopher C. *Antarctica and the Law of the Sea.* Dordrecht: Martinus Nijhoff, 1992.

Pickles, John. "Mapping the Geo-Body: State, Territory and Nation." Chap. 6 in his *A History of Spaces: Cartographic Reason, Mapping, and the Geocoded World.* London: Routledge, 2004.

Sambanis, Nicholas. "Partition as a Solution to Ethnic War: An Empirical Critique of the Theoretical Literature." *World Politics* 52 (2000): 437–83.

Dividing the Sea

Berret, Amy deGeneres. "UNCLOS III: Pollution Control in the Exclusive Economic Zone." *Louisiana Law Review* 55 (1995): 1165–89.

Boggs, S. Whittemore. "Problems of Water-Boundary Definition: Median Lines and International Boundaries through Territorial Waters." *Geographical Review* 27 (1939): 445–56.

Cook, Peter J., and Chris M. Carleton, eds. *Continental Shelf Limits: The Scientific and Legal Interface.* New York: Oxford University Press, 2000.

International Hydrographic Organization, Advisory Board on the Law of the Sea. *A Manual on Technical Aspects of the United Nations Convention on the Law of the Sea—1982.* Special Publication 51, 4th ed. Monaco: International Hydrographic Bureau, 2006.

Nurse, Leonard, and Rawleston Moore. "Critical Considerations for Future Action during the Second Commitment Period: A Small Islands' Perspective." *Natural Resources Forum* 31 (2007): 102–10.

Prescott, Victor, and Clive Schofield, eds. *The Maritime Political Boundaries of the World.* 2nd ed. Leiden: Martinus Nijhoff, 2005.

Rieser, Alison. "Essential Fish Habitat as a Basis for Marine Protected Areas in the U.S. Exclusive Economic Zone." *Bulletin of Marine Science* 66 (2000): 889–99.

Smith, Robert W. "The Maritime Boundaries of the United States." *Geographical Review* 71 (1981): 395–410.

Smith, Robert W., and J. Ashley Roach. *United States Responses to Excessive Maritime Claims.* 2nd ed. The Hague: Martinus Nijhoff, 1996.

Divide and Govern

Newman, David. "From the International to the Local in the Study and Representation of Boundaries: Theoretical and Methodological Comments." In *Holding the Line: Borders in a Global World,* ed. Heather N. Nicol and Ian Townsend-Gault, 400–413. Vancouver: UBC Press, 2005.

Smith, Gary Alden. *State and National Boundaries of the United States.* Jefferson, NC: McFarland and Co., 2004.

Stein, Mark. *How the States Got Their Shapes.* New York: HarperCollins, 2008.

Steinbauer, Paula E., and others. *An Assessment of Municipal Annexation in Georgia and the United States: A Search for Policy Guidance.* Athens: University of Georgia, Carl Vinson Institute of Government, 2002.

Contorted Boundaries, Wasted Votes

Amy, Douglas J. *Behind the Ballot Box: A Citizen's Guide to Voting Systems.* Westport, CN: Praeger, 2000.

———. *Real Choices/New Voices: How Proportional Representation Elections Could Revitalize American Democracy.* 2nd ed. New York: Columbia University Press, 2002.

Bowler, Shaun, Todd Donovan, and David Brockington. *Electoral Reform and Minority Representation: Local Experiments with Alternative Elections.* Columbus: Ohio State University Press, 2003.

Butler, Katharine Inglis. "Racial Fairness and Traditional Districting Standards: Observations on the Impact of the Voting Rights Act on Geographical Representation." *South Carolina Law Review* 57 (2006): 749–84.

Monmonier, Mark. *Bushmanders and Bullwinkles: How Politicians Manipulate Electronic Maps and Census Data to Win Elections.* Chicago: University of Chicago Press, 2001.

Redlining and Greenlining

Crossney, Kristen B., and David W. Bartelt. "The Legacy of the Home Owners' Loan Corporation." *Housing Policy Debate* 18 (2005): 547–74.

Hillier, Amy E. "Redlining and the Home Owners' Loan Corporation." *Journal of Urban History* 29 (2003): 394–420.

———. "Residential Security Maps and Neighborhood Appraisals: The Home Owners' Loan Corporation and the Case of Philadelphia." *Social Science History* 29 (2005): 207–33.

Peters, Alan, and Peter Fisher. "The Failures of Economic Development Incentives." *Journal of the American Planning Association* 70 (2004): 27–33.

Growth Management

Birch, Eugenie Ladner. "Practitioners and the Art of Planning." *Journal of Planning Education and Research* 20 (2001): 407–22.

Elliott, Donald L. *A Better Way to Zone: Ten Principles to Create More Livable Cities.* Washington, DC: Island Press, 2008.

Heuer, Tad. "Living History: How Homeowners in a New Local Historic District Negotiate Their Legal Obligations." *Yale Law Journal* 116 (2007): 768–822.

National Research Council, Committee on Characterization of Wetlands. *Wetlands: Characteristics and Boundaries.* Washington, DC: National Academy Press, 1995.

Seifert, Donna J. *Defining Boundaries for National Register Properties.* National Register Bulletin 21. Washington, DC: U.S. Dept. of the Interior, National Park Service, Interagency Resources Division, National Register of Historic Places, 1995.

Tiner, Ralph W. *In Search of Swampland: A Wetland Sourcebook and Field Guide.* New Brunswick, NJ: Rutgers University Press, 1998.

Vice Squad

Gerard, Jules B. *Local Regulation of Adult Businesses.* St. Paul, MN: Thomson/West, 2004.

Hubbard, Phil, and others. "Away from Prying Eyes? The Urban Geographies of 'Adult Entertainment.'" *Progress in Human Geography* 32 (2008): 363–81.

Kelly, Eric Damian, and Connie Cooper. *Everything You Always Wanted to Know about Regulating Sex Businesses.* Planning Advisory Service Report 495/496. Chicago: American Planning Association, 2000.

Lyons, Donald, F. Andrew Schoolmaster, and Paul Bobbitt. "Controlling the Location of Sexually Oriented Businesses (SOBs)." *Applied Geographic Studies* 3 (1999): 23–43.

Monmonier, Mark. "Keeping Track." Chap. 8 in his *Spying with Maps: Surveillance Technologies and the Future of Privacy.* Chicago: University of Chicago Press, 2002.

Yung, Corey Rayburn. "Banishment by a Thousand Laws: Residence Restrictions on Sex Offenders." *Washington University Law Quarterly* 85 (2007): 101–60.

No Dig, No Fly, No Go

Brown, Michael W. "Airspace Obstacles and TFR Trivia." *FAA Aviation News* 42 (November/December 2003): 1–8.

Ehrenberg, Ralph E. "Up in the Air in More Ways than One: The Emergence of Aeronautical Charts in the United States." In *Cartographies of Travel and Navigation*, ed. James R. Akerman, 207–59. Chicago: University of Chicago Press, 2006.

McIlmail, Timothy P. "No-Fly Zones: The Imposition and Enforcement of Air Exclusion Regimes over Bosnia and Iraq." *Loyola of Los Angeles International and Comparative Law Review* 17 (1994): 35–83.

Monmonier, Mark. *Cartographies of Danger: Mapping Hazards in America.* Chicago: University of Chicago Press, 1997.

Electronic Boundaries

Crampton, Jeremy W. "The Biopolitical Justification for Geosurveillance." *Geographical Review* 97 (2007): 389–403.

Dobson, Jerome E., and Peter F. Fisher. "Geoslavery." *IEEE Technology & Society Magazine* 22 (Spring 2003): 47–52.

LaMarca, Anthony, and Eyal de Lara. *Location Systems: An Introduction to the Technology behind Location Awareness.* San Rafael, CA: Morgan & Claypool, 2008.

Monmonier, Mark. "Keeping Track." Chap. 8 in his *Spying with Maps: Surveillance Technologies and the Future of Privacy.* Chicago: University of Chicago Press, 2002.

SOURCES OF
ILLUSTRATIONS

This list supplements figure captions with source notes for facsimile illustrations or geographic information used in compiling or drawing new artwork.

2.1: Alfred Cornell Mulford, *Boundaries and Landmarks* (New York: D. Van Nostrand Co., 1912), 16–17.

2.2: Alfred Cornell Mulford, *Boundaries and Landmarks* (New York: D. Van Nostrand Co., 1912), 24.

2.3: Alfred Cornell Mulford, *Boundaries and Landmarks* (New York: D. Van Nostrand Co., 1912), 26.

2.5: After F. J. Marschner, *Land Use and Its Patterns in the United States* (Washington, DC: Government Printing Office, 1958), 20.

2.8: U.S. Geological Survey, 1969, Manchester, OK-KS, 7.5-minute quadrangle map.

2.10: *Enos v. Casey Mountain, Inc.*, 532 So. 2d 703 (Fla. App. 5 Dist. 1988), map on 707.

2.11: Adapted from illustrations in U.S. Bureau of Land Management, *Public Lands Surveying: A Casebook Prepared by the Cadastral Training Staff* (Washington, DC, 1975), chapter D.

2.12: *Peuker v. Canter,* 63 P. 617 (1901), map on 671.

3.3: Courtesy NASA/Goddard Space Flight Center, Scientific Visualization Studio.

3.4: Compiled from James E. Hill, Jr., "El Chamizal: A Century-old Boundary Dispute," *Geographical Review* 55 (1965): 510–22, maps on 513 and 517; and "Convention between the United States of America and the United Mexican States," *International Legal Materials* 2 (1963): 874–81, map on 881.

3.5: (above) U.S. Geological Survey, 1955, El Paso, TX, 7.5-minute quadrangle map.

3.5: (below) U.S. Geological Survey, 1997, El Paso, TX, 7.5-minute quadrangle map.

3.6: Compiled from maps provided online by Israel's Ministry of Defence at www.securityfence.mod.gov.il.

3.7: U.S. Central Intelligence Agency, 1983, "West Bank and Vicinity," 1:150,000.

3.8: *Facts in Review* 3 (April 10, 1941), 182.

3.9: (left) U.S. Geological Survey, 1955, Vernon, NY, 7.5-minute quadrangle map,

3.9: (right) New York Department of Transportation, 1978, Vernon, NY, 7.5-minute quadrangle map.

4.1: Reduced and retouched from a hand-colored version in the Library of Congress's online Hispanic and Portuguese Collections Web site, http://www.loc.gov/rr/hispanic/guide/.

4.2: U.S. Library of Congress American Memory Web site, http://memory.loc.gov/ammem/.

4.3: E. Hertslet, *The Map of Africa by Treaty*, vol. 2, *Abyssinia to Great Britain and France*, 3rd ed. (London: His Majesty's Stationery Office, 1909), map facing 730.

4.4: Enlarged excerpt from "Map annexed to Agreement between Great Britain and France of May 24–July 19, 1906, sheet 1," in E. Hertslet, *The Map of Africa by Treaty*, vol. 2, *Abyssinia to Great Britain and France*, 3rd ed. (London: His Majesty's Stationery Office, 1909), map facing 848.

4.5: Isaiah Bowman, *The New World: Problems in Political Geography* (Yonkers-on-Hudson, NY: World Book Company, 1922), 4.

4.6: Ellen Churchill Semple, "Geographical Boundaries—II," *Bulletin of the American Geographical Society* 39 (1907): 449–63, map on 451.

4.7: Charles Seymour, "The End of an Empire: Remnants of Austria-Hungary," in *What Really Happened at Paris*, ed. Edward Mandell House and Charles Seymour (New York: Charles Scribner's Sons, 1921), 87–111, map on 104.

4.9: Instituto Geográfico Militar, *Atlas geográfico de la República Argentina*, 7th ed. (Buenos Aires, 1998), 96 and 23.

5.2: UN Commission on the Limits of the Continental Shelf, Web page for the Russian Federation's submission, http://www.un.org/Depts/los/clcs_new/submissions_files/rus01/RUS_CLCS_01_2001_LOS_2.jpg.

5.3: EEZ limit from Richard L. Price and Marc E. Vincent, "Map Supplement: Prospective Maritime Jurisdictions on the Polar Seas," *Annals of the Association of American Geographers* 73 (1983): 617–18.

5.4: Redrawn from map in U.S. Department of State, *Limits in the Sea* no. 91, 1980.

5.5: Modified from National Marine Fisheries Service map, www.fpir.noaa .gov/SFD/pdfs/2006_CNMI_PRIA_Compliance_Guide_1Dec2006 .pdf.

5.6: NOAA Northeast Regional Office Web site, http://www.nero.noaa.gov/ prot_res/porptrp/HPTRPNEClosure.pdf.

5.7: National Oceanic and Atmospheric Administration, 2005, Chart 13283, Portsmouth Harbor, 1:20,000.

6.1: U.S. Geological Survey, 1909, Frederick, MD, 15-minute quadrangle map.

6.2: Compiled from U.S. Census Bureau, *Geographic Areas Reference Manual*, 2005.

6.4: *Gomillion v. Lightfoot*, 364 U.S. 339 (1960), map on 348.

6.5: Compiled from detailed Census Bureau maps prepared for post-2000 redistricting and the Supreme Court map in figure 6.4. Because the maps were not fully compatible, the pre- and post-1957 boundaries are approximate.

6.7: (left) G. Etzel Pearcy, *A Thirty-Eight State U.S.A.* (Fullerton, CA: Plycon Press, 1973), 25.

6.7: (right) G. Etzel Pearcy, *A Thirty-Eight State U.S.A.* (Fullerton, CA: Plycon Press, 1973), 23.

6.8: Compiled from Lynn Perry, "The Circular Boundary of Delaware," *Civil Engineering* 4 (November 1934): 576–80; and State of Delaware, 2008, Newark West, 5-minute quadrangle map.

6.9: Compiled from various nautical charts; *New Jersey v. New York*, 526 U.S. 589 (1999), map at 601; and maps in Mark Stein, *How the States Got Their Shapes* (New York: HarperCollins, 2008), 188–91.

7.2: James Parton, *Caricature and Other Comic Art* (New York: Harper and Brothers, 1877), 316.

7.3: Mark Monmonier, *Bushmanders and Bullwinkles* (Chicago: University of Chicago Press, 2001), 3.

7.6: Redrawn by author from superdistricts maps created around 2005 and posted at FairVote Program for Representative Government, North Carolina, http://www.fairvote.org/media/research/superdistricts/ NorthCarolina.pdf.

7.7: Redrawn by author from superdistricts maps created around 2005 and posted at FairVote Program for Representative Government, Pennsylvania, http://www.fairvote.org/media/research/superdistricts/Penn sylvania.pdf.

8.1: Homer Hoyt, *The Structure and Growth of Residential Neighborhoods in*

American Cities (Washington, DC: Federal Housing Administration, 1939), 77.

8.2: Homer Hoyt, *The Structure and Growth of Residential Neighborhoods in American Cities* (Washington, DC: Federal Housing Administration, 1939), figures 20–24.

8.3: Syracuse Economic Growth Council, http://www.syracusecentral.com/.

9.1: New York Public Library, New York Zoning Maps, September 7, 2007, Maps: All Things About Maps and Spaces@NYPL, http://beta.nypl.org/blogs/maps/.

9.2: New York City, Department of City Planning, zoning map panel 8d, accurate through May 14, 2008, and posted online at http://www.nyc.gov/html/dcp/pdf/zone/map8b.pdf.

9.3: Jamesville Zoning Proposals, Town of DeWitt Zoning Amendment, Hamlet of Jamesville, Existing Zoning, draft April 8, 2008. Converted from color to graytone; labels added to identify categories.

9.4: Jamesville Zoning Proposals, Town of DeWitt Zoning Amendment, Hamlet of Jamesville, Proposed Zoning, draft April 8, 2008. Converted from color to graytone; labels added to identify categories.

9.5: Jamesville Zoning Proposals, Town of DeWitt Zoning Amendment, draft May 15, 2008, p. 5.

9.6: Donna J. Seifert, *Defining Boundaries for National Register Properties*, National Register Bulletin 21 (Washington, DC: U.S. Department of the Interior, National Park Service, Interagency Resources Division, National Register of Historic Places, 1995), 14.

10.1: Compiled by author from Town of Cicero municipal code and Onondaga County tax maps.

10.2: Onondaga County, NY, Office of Real Property Services, Town of Cicero Tax Map no. 065; online at http://ocfintax.ongov.net/Imate/taxmaps.aspx.

10.3: Boston Redevelopment Authority, Leather District and South Station EDA, Chinatown, Bay Village Neighborhood Districts map sheet, http://www.cityofboston.gov/bra/pdf/ZoningCode/Maps/1CGN.pdf.

10.6: Arlington, Texas, Police Department, www.arlingtonpd.org/sexoffen/ordinance/SexOffenderBufferMapNorthDistrict.pdf.

11.1: Federal Aviation Administration, Baltimore-Washington Charted VFR Flyway Planning Chart, 76th ed., 1:250,000, February 14, 2008. Excerpt enlarged to 169 percent; area shown is approximately nine miles from left to right. Original in color.

11.3: Office of Coast Survey Historic Map and Chart Image Catalog, NOAA Central Library, http://historicalcharts.noaa.gov/historical_jpgs/J_18-8-19.

11.4: Thoburn C. Lyon, *Practical Air Navigation and the Use of the Aeronautical Charts of the U.S. Coast and Geodetic Survey*, Special Publication no. 197, 2nd ed. (Washington, DC: Government Printing Office, 1938), 15.

11.5: *Air Commerce Bulletin* 6, no. 11 (May 15, 1935): 268.

11.6: Federal Aviation Administration, "D.C. Metropolitan ADIZ and FRZ," NOTAM (Notice to Airmen), Effective 0500 UTC August 30, 2007, p. 11.

11.7: Official U.S. Air Force "map/infographic" online at www.af.mil/shared/media/ggallery/hires.afg_021215_005.jpg (retrieved October 9, 2008).

11.8: *Department of State Bulletin* 26 (March 24, 1952): map facing p. 451.

11.9: J. L. De Witt, *Final Report: Japanese Evacuation from the West Coast, 1942* (Washington, DC: Government Printing Office, 1943), 98.

11.10: Courtesy New York State Department of Environmental Conservation, which provided a high-resolution color image and approved modifications for this black-and-white reproduction.

11.11: California Department of Food and Agriculture, Fruit Fly Treatment Maps, http://www.cdfa.ca.gov/phpps/pdep/treatment/treatment_maps.html.

12.2: Michael Zuschlag, *Potential Interventions by Government and Industry to Minimize Violations of Temporary Flight Restrictions,* report NASA/CR—2005-213924, NASA Scientific and Technical Information Program Office (September 2005), figure 6 on 19.

12.3: Captured from www.google.com/maps on December 22, 2008.

INDEX

Page numbers in italics refer to figures.

Aberdeen Proving Grounds, 166, *167*

Abrams, Charles, 118

access easement, 28–29

accreted land, apportioning of, 26–27

accretion, 26

address-matching software, 156

Adélie Land (French Antarctic claim), 65–66

administrative boundaries, hierarchies of, 89, 101–3, 190n1

adverse possession, 24–25

aerial photos, 136

aeronautical chart: diverse formats, 166; electronic, 184; hazards shown, 165–66; as prohibitive cartography, 2, 162–70; symbols, 165–66, 168, 212n7

Africa: artificial boundaries, 51; colonial mapping of, 54–58; decolonization, 58; as obstacle between Europe and India, 54–55; partition of, 4, 55, 58

African Americans, voting rights of, 108, 114, 203n15

Air Commerce Act, 166, 169

Air Commerce Bulletin, 168–69

Air Defense Identification Zones (ADIZs), 169–70, *171*, 212n17

airspace, cartographic production of, 162–70

Aksai Chin, 35

Alabama, municipal annexation in, 94, 95

Alaska: air defense zones, 212n17; county equivalents, 89; maritime boundaries, 76–77; rectangular survey system, 17

Alexander VI (pope), 52

Alpha-Mendeleev Ridge, 75

Amarillo, TX, 111

American City, The, 131

American Geographical Society, 59

American Planning Association, 147, 148

American Samoa, 80–81

Amundsen, Roald, 67

ancient landmark (in property surveying), 12

Andorra, 46

Angola, 58

Antarctica: research stations, 69; and sea level rise, 81; territorial claims to, 64–69; treaty, 68–69, 195n33

Antarctic Treaty System, 68–69
Antártida Argentina, 66
appurtenant easement, 23
Arafat, Yasir, 44
architectural restrictions (in land-use zoning), 139–43
Arctic Ocean, 76–77
Argentina: Antarctic claim, 64–67, 195n23; border with Chile, 41; rhetorical cartography, 33–34, 66, 67
Arlington, TX, residency restrictions on sex offenders, 156, *157*
Atafu (Tokelau), 81
Atlanta, GA, 174
Atlantic Monthly, 43
Atlas of England and Wales, 88
Atlas of Historical County Boundaries, 95, 201n26
Australia: air defense zones, 169; Antarctic claim, 64–67; proportional representation in, 112
Austria-Hungary, 62–63
avulsion, 26
Azores, 52

Balkans, 51, 59, 63
Ball, Todd, 145
Baltimore, MD, alignment of street grid, 12
Barents Sea, 75
base line (element on Public Land Survey), 14–18
baseline, for maritime boundary, 71–72
Bay Shore Historic District (Miami, FL), 142
beachfront, as property boundary, 26, 29
Beacon Hill (Boston), 151
bearing: ancient, 12; in metes and bounds surveying, 11–12
Beaufort Sea, 76, 77
Beichman, Arnold, 173
Belgium, colonial territory of, 55
Benin. *See* Dahomey
Bering Strait, 76, 77
Berke, Richard L., 202n1
Berkeley, CA, 175
Berlin Conference of 1884, 55
Berlin Wall, 43

Big Diomede Island, 76
Big Sur, CA, 164
Birmingham, AL, 94
Bismarck, Otto von, 55
Bissinger, Buzz, 119
Blaine, WA, 36–37
block census data, in congressional redistricting, 109
block number, 7, 19
Bloodsworth Island, MD, 163
Board of Surveys and Maps, 166
Board on Air-Space Reservations, 167–68
Bohai Bay, 72
borrowed boundaries, ZIP codes as, 123–24
Bosnia and Herzegovina: disintegration of Yugoslavia, 63; no-fly zone, 172
Boston Gazette, 108
boundaries: ambulatory, 4, 25–29, 38–39, *40*; artificial, 42, 51, 57; astronomical, 51, 56; barriers to movement, 30; block and lot systems, 7, 18–22; circular, 19–20, 98–100; colonial, 4, 51–58; county, 95–96; culture as basis for, 41–42; defensible, 42–43; delimitation of, 39–41, 50, 55; demarcation of, 42, 50, 55–56; drainage divide as, 41, 56; ethnicity and, 59–64; geometric, 56; geopolitical stability, 51; hierarchy of, 89, 136; of historic preservation districts, 2, 141–43, 208n18; international, 30–50, 190n1; for land-use regulation, 136–41; maritime, 70–85; markers for, 9–10, 12, 20; national, 3; natural, 42; offset of, 12, 57–58; political protest exclusion zones, 181, 214n3; race as basis for, 41–42, 92; rivers as, 4, 26–28, 56, 100–101, *102*; scientific objectivity and, 51; shoreline, 26, 29; state, 86, 96–101, 190n1; straight lines as, 7, 13, 41, 75–77; submerged land, 101, *102*; turning points for, 78, 83; of voting districts, 104–17, 185; for wetlands restrictions, 143–45
boundary commission, role of, 56
boundary lines on maps: electronic representation of, 88; length of, 11–12; in residential subdivisions, 19–21; rhetorical role, 2, 30–34, 129; social acquies-

cence to, xii, 2, 185–86; symbols for, 2–3; on zoning maps, 136
boundary stones, 3, 9
Bowman, Isaiah, 59–63
Brazil: Antarctic aspirations, 68; discovery of, 52; Tordesillas Line and, 53, *54*
Britain: Antarctic claim, 64–67; colonial territory of, 53–58; economic redevelopment zones, 205n30; no-fly zones over Iraq, 171; origin of metes and bounds surveying in North America, 7; at Paris Peace Conference, 59; resistance to expanded territorial seas, 73; sailors seized by Iran, 83
Bryant Park (New York City), 134
Buckingham Palace, 169
buffer zone, to restrict sex offenders, 156–58
building envelope (in land-use zoning), 132
building height, regulation of, 132
Bullwinkle District, 109
Bureau of Indian Affairs, 49, 50, 103
Bureau of the Census, 88–89, 94
buried infrastructure. *See* underground facilities
Bush, George Herbert Walker, 109, 165
Bush, George Walker, 36, 165, 186
Bushmanders, 109
Bustamante, Marcos, 68

cable, buried. *See* underground facilities
Cabot, John, 53
California: municipal annexation in, 95; rectangular survey system, 17; redlining for auto insurance banned, 124
California Department of Food and Agriculture, 179
call (part of parcel perimeter), 8
Cambridge, MA, 112–13, 203nn9–10
Camden County, NJ, 174
Camp David, 164–65
Canada: air defense zones, 169; land boundary with United States, 35–37; maritime boundary with United States, 75–77, 83, 197n21
Cantino, Alberto, 53
Canton, OH, 155
Cape Cod, 83
Cape Verde Islands, 52

Capitol (building), 162, 169
census geography, 89, 109
Census of Governments, 90
Center for Voting and Democracy. *See* FairVote
Central Europe: ethnic regions, 59–62; international boundaries, 41
Central Intelligence Agency (CIA): map of Antarctica, 68; map of West Bank, 43–44, *45*; World Factbook, 81
chain: length of, 11; measuring instrument, 11–12; reliability, 12, 18
Chamizal Convention, 38–39, *40*
Chamizal National Memorial, 39, *40*
Charlotte, NC, regulation of sex-related businesses, 151–52
Chehaw, AL, 94
Cheney, Richard, 164, 186
Chesapeake Bay, 166, *167*
Chicago, IL, 95, 165
Chicago school (sociology) theory of urban ecology, 120–21, 129
Chile: Antarctic claim, 64–66, 68; border with Argentina, 41; maritime boundary, 72, 73, 82
China: boundary with India, 31, 35; early military maps, 88
choice voting, 112–13, 115, 203nn9–10, 203n16
Chrysler Building, 132
Chukchi Sea, 76, *77*
Cicero, NY: regulation of sex-related businesses, 149–50, *151*; residency restrictions on sex offenders, 156
circular arc, as part of boundary, 20, 56, 98–100
City Survey Program, 118–19
Clean Water Act, 144–45
closing line (for maritime boundary), 72, 78
Code of Federal Regulations, 84–85
Cold War, 172–74, 180
collision avoidance, 183–85
colonial America, land subdivision in, 12–14
colonization, 45, 51–58
Colorado: municipal annexation law, 93; rectangular survey system, 17
color symbols (on prohibitive maps), 2, 118, 119, 124, 132, 166, 168

Columbus, Christopher, 52, 53
Columbus, NM, 37
Combat Zone (Boston), 150–51, *152*
Commission of Inquiry (U.S.). *See* Inquiry, The
compactness, in political redistricting, 106, 110
compass (surveyor's), reliability of, 12
Connecticut, closed to Russian diplomats, 174
Consumer Federation of America, 124
contiguity, in municipal annexation, 92–95
contiguous zone, 73
continental shelf, 74–76
Convention on the Territorial Sea and Contiguous Zone, 73
Cook Islands, 80
Cooper, Connie, 148–49
Cordova Island, 38–39, *40*
corner points: in metes and bounds survey, 8, 9; in Public Land Survey, 17–18; in residential subdivision, 7, 19–20
corner siting (of locally objectionable land uses), 149
Council of State Governments, 155–56
counter map, 185–86
county boundaries, 95–96
county census division (CCD), 89
county subdivisions (in United States), 88–90
cracking (form of gerrymandering), 109–10
Crawford, TX, 165
Croatia, 63
Croats, 63
cumulative voting, 112
Cvijić, Jovan, 63
Cyprus, boundary between Greek and Turkish factions, 35
Czechoslovakia, 62–63
Czechs, 60–63

Dahomey, 58
Daily Mail, 31, *32*
Danbury, CT, locational restrictions on sex offenders, 158
Darfur, 4
decolonization, 58

Delaware: boundary, 98–100; closed to Russian diplomats, 174
DeLong, James, 144
Demko, George, 58
Democratic Party, 104, 108–9, 114
Democratic Republic of Congo, 58
Denton, Nancy, 119
Department of Commerce, 167, 169
Department of Homeland Security, 37
Department of Justice, 108–9
Department of State, 75–76, 97, 172–74
Department of the Treasury, 16
Department of War, 166, 167
Detroit, MI, 91–92
DeWitt, NY: historic preservation, 140–41; land-use zoning, 136–41; regulation of sex-related businesses, 148–49, 150; sign ordinance, 137–38
discovery principle (for territorial claims), 51–52
Disneyland, 164
distance-based restrictions: on sex offenders, 155–59; for sex-related businesses, 147–48, 150–52
Dobson, Jerry, 181
dominant tenement, 23
Dominican Republic, 34, *35*, 78
Dumont d'Urville, Jules Sébastien César, 65

easements, 4, 7, 23–25
East River (New York City), 96
economic redevelopment, as motive for greenlining, 124–29
Ecuador: Antarctic aspirations, 68; boundary with Peru, 31
Elliott, Donald, 132
Ellis Island, 101, *102*
El Paso, TX, 38–39, *40*
El Salvador, 73
Empire Zone (EZ), 126–28, 206n37
empowerment zone, 124–25
encroachments, 4, 6–7, 23–24
England. *See* Britain
Enos v. Casey Mountain, 24–25
enterprise zone, 124–25, 205n30
environmental maps, as prohibitive cartography, 1, 143–45, 175, 177

equidistance principle (for maritime boundaries), 76, 78–80
Equitable Building (early New York City skyscraper), 132, 207n6
erasures, cartographic, 33
erosion, 26–28
Ethiopia, 55
ethnic cleansing, 63
Euclid, OH, 131
Eurafrica, 46
European Union (EU), cross-boundary initiatives, 31
exclusion maps, 174–75, *176*
exclusive economic zone (EEZ), 70–83

Facts in Review, 46
FairVote, 113–15
Fakaofo (Tokelau), 81
Falkland Islands, 33–34, 64–66
fault-zone map, 1, *177*
Federal Aviation Administration (FAA), 162–64, 169
Federal Geographic Data Committee, 88
Federal Housing Administration (FHA), 118, 121, 123, 129, 204n3
federal land trust, 50, 102–3
Federal Register, 162
Fiscal Policy Institute, 126
Fisher, Peter (geographer), 181
Fisher, Peter S. (economist), 125
Fisheries Conservation Zone, 74
fishing rights, 71, 73–74, 77, 82
Flight Service Station (FSS) briefings, 184
floodplain map, 1, 143, 177, 208n26
floor-area ratio (FAR), 135
Florida, municipal annexation in, 92, 95
forbidden cities, 172–74
Ford, George, 131
France: Antarctic claim, 64–66; colonial territory, 55, 56; exclusive economic zone, 78; no-fly zones over Iraq, 171; at Paris Peace Conference, 59
Frankfurter, Felix, 92
French Soudan, 57–58
fruit fly infestation, 178–79
full representation, 111–12
full-value assessment, 22

Galápagos Islands, 68
Gambia, 56
Garment District (New York City), 134
General Land Office (GLO), 16–18, 27, 49
Geneva, NY, 127
gentrification, 143
geodesic line, 78
Geographical Review, 63
geographic information systems (GIS): buffer restrictions on sex offenders, 156, 158; location tracking and, 180; for underground facilities, 161–62
Georges Banks, 83
Georgetown (Washington, DC), 141
Georgia: municipal annexation in, 95; on-line sex-offender registry, 153–54
Georgia Bureau of Investigation, 154
GeorgiaSexOffenders.com, 154
German Library of Information, 46
Germany: Antarctic aspirations, 67–68; colonial territories, 55, 59; ethnic outliers, 60–61; geopolitical mapping by, 46, *47*; protested map on cover of *L'Illustration*, 32–33
Gerry, Elbridge, 108, 109
gerrymander: hypothetical illustrations, 105–7, 110–11; nineteenth-century prototype, 104, 107–8; origin of term, 107–8; partisan (political), 104–11; as prohibitive cartography, 104–5, 117
Ghana, 58
gill nets, 83
Gold Coast, 57–58
Google Maps, 154, 186–87
Government Accountability Office, 102, 125
GPS (global positioning system): and air traffic control, 183; and buffer restrictions on sex offenders, 158; limitations of, 182–83; locating underground facilities with, 4, 161; location tracking and, 180; and maritime territory, 71; property mapping with, 22; vehicle navigation, 180
Graham Land, 64
Grand Central Station, 133
Grand Cypress Resort, 24–25
grandfathering, 139

Great Wall of China, 42
Greenbaum, Robert, 125
Greenland, 81
Green Line: on Cyprus, 35; between Israel and Jordan, 43
greenlining: compared with redlining, 128–29; for distributing benefits and liabilities, 128; economic redevelopment and, 117, 124–28; effectiveness questioned, 125–26, 128; map as distraction, 128–29; origin of term, 124
grievance day, 22
gross easement, 23
ground-penetrating radar, 161
Guam, air defense zones, 212n17
Guano Islands Act, 80
Gulf of Maine, 83
Gulf of Sidra, 72
Gulf War (1991), 171

Haakon VII (king of Norway), 66–67
Hadrian's Wall, 42
Hagerstown, MD, 164, *165*
Haiti, 34, *35*
Hamtramck, MI, 91–92
Harney, Kenneth, 124
Hawaii: air defense zones, 212n17; maritime boundary, 80, 97; municipal annexations, 95; shoreline, 26
hazardous environments, as impetus for restrictive cartography, 2, 143
Herrera y Tordesillas, Antonio de, 53, *54*
Highland Park, MI, 91–92
High Seas, 71, 80
Highway Beautification Act, 103
Hillier, Amy, 122–23
Hills, Edmund, 56–57
Hispanic Americans, voting rights of, 108–9
historic preservation, 2, 140–43
Hitler, Adolf, 42
Holdich, Thomas Hungerford, 41–42, 45
Home Loan Bank Board, 118
Home Owners Loan Corporation (HOLC), 118–23, 124, 129, 204n3
Homestead Act of 1862, 14
Hoyt, Homer, 121–22
Hudson River, 101, *102*

Hulhumalé (artificial island in Maldives), 81
Hungarians. *See* Magyars
hunters, regulatory maps for, 160, 175, 177
Huske, John, 54, *55*
Hussein, Saddam, 63–64, 171

Illinois, 95
imperative map. *See* prohibitive cartography
independent redistricting commissions, 111
India: Antarctic aspirations, 69; boundary dispute with Pakistan, 31, 35; embargo of Microsoft Windows 95 operating system, 30–31
Indian land, tax exemption of, 49
Indiana, 95
Indian Nonintercourse Act of 1790, 47
initial point, 14, *15*
innocent passage, 72
Inquiry, The, 59–63
instant-runoff voting, 203n16
Intergovernmental Panel on Climate Change, 81
internal waters, 72
International Boundary and Water Commission (Mexico and United States), 37
International Boundary Commission (Canada and United States), 36
International Court of Justice, 43, 82, 83
International Geophysical Year, 68
International Seabed Authority, 74, 75
Internet: electronic mapping on, 186–87; flight restrictions posted on, 164, 169–70; hunting regulations posted on, 175, 177; pornography, 151–52; sex-offender registries, 153–56
Iran, 83
Iraq, 35, 63, 83, 170–71
Ireland, proportional representation in, 111
Isla Aves (Venezuela), 78
islands: artificial, 81–82; and exclusive economic zone, 71, 77–81; and maritime boundaries, 71; sea level rise and, 71, 81–82; and territorial sea, 71, 81–82
Islas Malvinas. *See* Falkland Islands
Israel: boundary with Jordan, 43; place-

names to claim territory, 44–45; Security Fence, 43–44, 192n33; West Bank encroachments, 43–45
Italy, colonial territory of, 55

Jackson, Kenneth, 119
Jamesville (hamlet), NY: greenlining in, 127–28; historic preservation, 140–41; land-use regulation, 136–41
Japan: air defense zones, 169; maritime boundary with United States, 70; whaling, 68
Japanese Americans, wartime exclusion zones for, 174–75, *176*
Jefferson, Thomas, 14
Jennings, Eli Hutchinson, Sr., 81
Jerusalem, 43
John II (king of Portugal), 52
Johnson, Lyndon, 118
Johnstown, PA, 174
Jones, Barclay, 125
Juárez, Mexico, 38–39
Jugoslavia, 62–63
jurisdictions, types of, 86

Kansas: municipal annexation law, 95; rectangular survey system, 17
Kashmir, boundary dispute, 31, 35
Kelly, Eric, 148–49
Kennebunkport, ME, 165
Kennedy, John F., 38
Kilauea, HI, 164
Kingdom of Serbs, Croats, and Slovenes, 63
Kings County (Brooklyn), NY, 96
Kittery, Maine, 84–85
Korea, Demilitarized Zone, 34–35, 43
Kurds, 64, 171
Kuwait: boundary with Saudi Arabia, 35; Gulf War, 171

Lakewood, CA, *178*
landlocked countries, 46
landlocked lot, 23
landmarks, property boundaries and, 8–12
Land Ordinance of 1785, 14
land ownership, 6, 185
land registration system, electronic, 21–22

land survey: easements and encroachments, 23; mentioned, 6–7; metes and bounds, 7–14; patented, 18; retracement and resurvey, 18, 29; for subdividing federal lands, 14–18; for subdividing private parcels, 18–21
land use, regulation of, 130–45
land-use map, 131
land-use restrictions, 130–52, 185
Law of the Sea, 73–83
lebensraum (living space), 42
Leu, Shirley-Ann and Herbert, 36
Liberia, 55
Library of Congress, 108
Libya, 72
L'Illustration, 31–33
limited voting, 112
Lincoln Memorial, 169
line tree, 9
Little Creek Naval Amphibious School, 164
Little Diomede Island, 76
Livermore, NH, 89–90
location-based services (LBS), 180–82
location histories, 158
location tracking: abuse of, 180–81; as alternative to incarceration, 182, 213n2; in automobile insurance, 181; in automobile rental, 181; reliability of, 182–83; social acquiescence to, 180, 185–86; technologies required, 180; for transportation safety, 183–85; unintended consequences, 182
Long, John, 95–96
longitude, accuracy of, 53
"long lasso" annexation, 92–94
lopped tree, 9, *10*
Los Angeles, CA, 95
lot number, 7, 13, 19
Louisiana: counties in, 89; exclaves of, 97; waterfront boundaries, 26
Lowe's, 126
low-water shoreline, 71

Macedonia, 63
Macon County, LA, 92
Madison, James, 108
Madison County, NY, 47–48

Magellan, Ferdinand, 64
Maginot Line, 43
magnetic declination, 12
Magnuson Fishery Conservation and
    Management Act, 74
Magyars, 60–63
Maine, 90, 200n15
Malacca, 53
Malaysia: Antarctic aspirations, 69
Maldives, 81–82
Male (Maldives), 81
Malone, James L., 74
Manihiki (Cook Islands), 80
Manlius, NY, 149–50
Manson, Donald, 125
mapping, changing roles of, 2
MapQuest, 153
marine managed area. *See* marine pro-
    tected area
marine protected area, 2, 71, 83–84, 199n46
maritime territory: closing lines, 72, 78;
    contiguous zone, 73; continental shelf,
    74–76; equidistance principle, 76, 78–
    80; exclusive economic zone (EEZ),
    70–83; historic claims, 72; seaward ex-
    tension of boundary, 76–77; submarine
    topography and, 74–76; territorial sea,
    71–73, 81–82
Marschner, Francis, 13
Maryland: state boundary, 98–100; sub-
    county units, 89
Mason-Dixon Line, 89, 90, 98
Massachusetts: original gerrymander, 107–
    8; closed to Russian diplomats, 174; law
    on waterfront boundaries, 26
Massey, Douglas, 119
master plan, 131
McKinley, William, 155
meander line, 28
Megan's Law, 153
Mercator, Gerard, 64
meridians, as boundaries, 52–53
metes and bounds: defined, 7–8; in New
    England, 12–14; origin, 7; surveying
    of, 7–12
metropolitan statistical area (MSA), 102
Mexican border. See United States–
    Mexico border

Microsoft Windows 95, 30–31
microstates, 46
Middle East: artificial boundaries, 51;
    international boundaries, 41, 43–46;
    post–World War I mandates in, 59
Midtown Manhattan, zoning maps for,
    132–36
Minnesota Department of Corrections, 157
minor civil division (MCD), 89–90
minority-majority congressional districts,
    108–9
Mississippi, 97
Molucca Islands, 53
Monroe, James, 47
Montenegrins, 63
Montenegro, 63
monuments: for land subdivision, 20–21;
    for municipal survey system, 21; offset
    of, 21
Mount Ephraim, NJ, 147, 149
Mount Vernon, 163
Mulford, Alfred, 8, 9, 12
multimember districts: constitutionality
    of, 203n15; with proportional repre-
    sentation, 112–16; proposed for North
    Carolina, 114, 115; proposed for Penn-
    sylvania, 114–15; weighted voting in
    House of Representatives, 116
municipal boundaries: annexation and,
    91–95, 200n15; impact of, 86–87; in-
    corporation and, 91; prohibitive car-
    tography and, 2
Murmansk, 75
Murphy, Robert, 32, *33*
Murray Hill (New York City neighbor-
    hood), 133

National Aeronautics and Space Admin-
    istration (NASA), 184
National Archives, 119
national boundaries, rise of, 3
National Geospatial-Intelligence Agency,
    186–87
National Imaging and Mapping Agency.
    See National Geospatial-Intelligence
    Agency
National List of Vascular Plant Species
    That Occur in Wetlands, 144, 208n31

National Marine Fisheries Service, 83
National Oceanic and Atmospheric Administration (NOAA), 83–84
National Park Service, 141
National Register of Historic Places, 141
National Security Areas, 164
National Spatial Data Infrastructure, 88
Nauru, 46
nautical chart: electronic, 78; evidence in boundary dispute, 101; plotting location on, 78; as prohibitive cartography, 2, 83–85; restrictions noted on, 84–85
Naval Observatory, 162–63, 186
Nebraska, rectangular survey system for, 17
Netherlands Antilles, 78
Neuschwabenland, 68
Nevada: municipal annexation law, 95; rectangular survey system, 17
New Castle, DE, 99–100
New England: land subdivision in, 12–14; local governments, 90; municipal annexation in, 200n15
"New England Extended," 90
New Guinea, 46
New Hampshire, boundary, 89–90; wildlife conservation areas, 175
New Haven, CT, 119
New Jersey: boundary with Delaware, 100–101; municipal annexation in, 95, 200n15
New York City: land-use zoning, 131–36; submerged lands, 101–2; and travel by Soviet diplomats, 173–74
New York State: child safety zones, 156; Department of Environmental Conservation, deer-hunting restrictions, 175, 177; Department of Transportation, cartographic treatment of Oneida Nation Territory, 48–49; Empire Zone (EZ) program, 126–28, 206n37; local governments, 88; sex-offender registration, 153; submerged land boundary, 101, 102; waterfront boundaries, 26, 101, 102; wildlife management units (WMUs), 175, 177
New York Times, 174, 202n1
New Zealand: Antarctic claim, 64–66;

maritime boundary with United States, 70, 80–81
Nigeria: civil war, 4; colonial boundaries, 58
no-fly zone, 2, 166, 170–72
noncontiguous annexation, 95
North Carolina: multimember congressional districts proposed for, 114, 115; municipal annexations in, 95
North Korea. See Korea, Demilitarized Zone
North Pole, 76
Norway, Antarctic claim, 64–67
Notices to Airmen (NOTAM), 162, 163–64, 184
NRG Energy (firm), 126
Nukunonu (Tokelau), 81

Oakland, CA, 175
Obama, Barack, 165
Office of Indian Affairs. See Bureau of Indian Affairs
O'Hare Airport, 95
Ohio: online sex-offender registry, 154–55; rectangular survey system for, 17
Oklahoma, municipal annexation laws, 92–93
Olohega (American Samoa), 81
Oman, boundary with Saudi Arabia, 35
one-call centers, 161–62
Oneida County, NY, 174
Oneida Nation Territory, 47–50, 103
Onondaga County, NY, 96, 127
Onondaga Indian Reservation, 49
Ontario County, NY, 127
Orwell, George, 183, 187
Otsego County, NY, locational restrictions on sex offenders, 157–58
Ottoman Empire, 59
overlay analysis: in redistricting, 109; redlining and, 121, 122–23
oxbow lake, 26

packing (form of gerrymandering), 106–10
Pakistan, boundary dispute with India, 31, 35
Palestine, controversial maps of, 33, 43–45
Palestinian Authority, 44

paper shrinkage, 19
Paris Peace Conference, 58–63, 68
Park, Robert, 120–21, 129
Parrott, James, 126
Patagonia, 41
Pearcy, G. Etzel, 96–98, 99
pedophiles. *See* sex offenders
Penn, William, 99
Pennsylvania: boundaries 98–99, 101; multimember congressional districts proposed for, 114–15; municipal annexation in, 200n15
Penrhyn (Cook Islands), 80
Pentagon, 170, *171*, 187
Peoria, IL, 111
perpendicular allocation, of accreted land, 27
Peru: Antarctic aspirations, 68; boundary with Ecuador, 31; territorial sea, 73
Peter I (king of Serbia), 63
Peters, Alan, 125
petroleum exploration, 71, 76, 77
Peuker v. Canter, 27–28
Philadelphia: Delaware River and, 98, 101; redlining in, 118–20, 122–23; travel restrictions for Russian diplomats, 174
photomaps, 22
pipeline. *See* underground facilities
place-based discrimination: in automobile insurance, 124; in automobile rental, 123; for economic redevelopment, 124–29; in home mortgages, 117–24; and performance-based discrimination, 129; racial overtones, 118, 119, 123
place names, in Israel and Palestine, 43–45
plat: in rectangular land survey, 13–15, 18; for residential subdivision, 19–20
point of beginning, 8
Poland, 59–60
Political Frontiers and Boundary Making (Holdich), 41–42
Portsmouth Naval Shipyard, 85
Portugal, colonial territory of, 52–53, *54*, 55
positive train control, 184
Potsdam, NY, 127, 206n46
preference voting, 112
prescriptive easement, 24

principal meridian, 14–17
prohibited areas (on aeronautical charts), 162–63, 166–68, 211n5
prohibitive cartography: electronic, 180–87; gerrymandering as, 104–17; greenlining as reverse form of, 124–29; as humanitarian intervention, 2, 170–72; for land-use regulation, 130–52; map symbols for, 2–3; momentum of, 187; and political protest, 181, 214n3; prediction as source of power, 117; redlining as, 117–24, 128–29; for sex offenders, 152–59, 180; social acquiescence to, xii, 2, 180, 185–86; threats to privacy, 180, 185–87; for transportation safety, 183–85; travel restrictions and, 172–74; as twentieth-century phenomenon, xii, 1–2, 186; underlying factors, 2; validation of government control, 185; for wildlife conservation, 2, 175, 177
property mapping: Australia, 22; Babylonia, 7; Canada, 22; Egypt, 3, 7; England, 7; GPS and, 22; land ownership and, 6, 29, 185; land survey and, 6–7, 23; roles of, 29; Rome (ancient), 1–2, 3, 7; tax assessment and, 21–22; Western Europe, 22
proportional representation: advantages, 111–12, 116; in Cambridge, Massachusetts, 112–13; in multimember districts, 111–16; objections to, 115; single transferable vote, 112–13, 203n9–10; as virtual redistricting, 111; and weighted voting in House of Representatives, 116
proportionate allocation, of accreted land, 26–27
protective incorporation, 91
public administration, as impetus for restrictive cartography, 2
Public Land Survey (United States): accuracy, 18; adjustment for inequality, 15–17; checkerboard landscape, 16–17; elements of, 14–18; patenting of lines and corners, 18; plats, 18; resurvey for, 18; zones for, 17
Pukapuka (Cook Islands), 80

quarantine, agricultural, 178–79
Quarryside Animal Hospital, 127, *128*
quarter corner, 18
Queen Maude Land, 67–68
Queens County, NY, 96

Rae, Douglas, 119
Rakahanga (Cook Islands), 80
RAND Corporation, 186–87
Rand McNally, 166
range (element of U.S. Public Land Survey), 15–16
range line, 14–16, *16*
Reagan, Ronald, 73–74
Reagan National Airport, 170, *171*
real estate. *See* real property
real property: block and lot surveys, 19–21; legal description, 4, 9, 22; tax assessment, 21–22
rectangular survey, 13. *See also* Public Land Survey (United States)
red-light district, 149
redlining: in automobile insurance, 124; in automobile rental, 123; for distributing benefits and liabilities, 128; compared with greenlining, 128–29; for home mortgages, 118–24; origin of term, 118; overlay analysis for, 121–22
Reliant Energy (firm), 127
Republican Party, 104, 108, 114
residential land, subdivision of, 18–22
Residential Security Maps, 119
restricted areas (on aeronautical charts), 162–67, 183–84, 211n5
restrictive declarations (in land-use zoning), 135–36
restrictive map. *See* prohibitive cartography
retracement survey, 18
Reynaud, Paul, 32–33
rhetoric, graphic, 2, 30–31, 51–54, 129
Rhode Island, closed to Russian diplomats, 174
Richmond, VA, 121, *122*
right-of-way, 4, 7, 23–25
Rio Bravo del Norte. *See* Rio Grande
Rio Grande, 37, *39–40*
riparian rights, 26–29

rivers, as boundaries, 4, 26–28, 100–101, *102*
Romania, 62
Roosevelt, Franklin, 168–69, 174–75
Royal Geographical Society, 41
rural, problematic definition, 102
Rural Housing Service, in Department of Agriculture, 102
Russia: Antarctic claims rejected by, 68; Arctic claims, 66–67, 75–77; following World War I, 59–60; travel restrictions on American diplomats, 172–74

Salina, NY, 156
San Marino, 46
satellite positioning. *See* GPS (global positioning system)
Saudi Arabia, boundaries with neighboring countries, 35
Saxton, Christopher, 88
Saxton McKinley House (Canton, OH), 155
scenic easement, 23
Schad v. the Borough of Mount Ephraim, 147
school attendance zones, 3, 87
Scott, Roger, 130
seabed mining, 71, 74
seafloor mapping, 74
sea level rise, 71, 81–82
Secret Service, 163
sections: corners of, 17–18; element of U.S. Public Land Survey, 14–16; numbering of, 14, *15*; subdivision of, 18
sector principle, 65–69, 75–77
security supporting airspace (SSA), 184
Seminole, OK, 93
Semple, Ellen Churchill, 60–62
sense of place, 91
Seoul, Korea, 34
Serbia, 63
servient tenement, 23
Seward, William, 76
sex offenders: location tracking of, 180, 213n2; recidivism, 152–53, 159; online registries, 153–56; residency restrictions, 4, 155–59
sex-related businesses, regulation of, 146–52

shape. *See* compactness, in political redistricting

Shiite Arabs, 64, 171

shirt changer, 126

Shivwits Band of Paiute Indians, 103

Siachen Glacier, 35

signage, regulation of, 103, 135, 137–38

Silicon Valley, 174

Six Nations of the Iroquois, 47

Skaneateles, NY, 130

Slovaks, 60–63

Slovenia, 63

Smithsonian Atlas of World Aviation, 162

Society Hill (Philadelphia), 141

soft partition, 63–64

South Africa, Antarctic aspirations, 68

South Georgia Islands, 66

South Korea. *See* Korea, Demilitarized Zone

South Orkney Islands, 66

South Pole, anchor for sector claims, 64–67

South Sandwich Islands, 65–66

Sovereign Military Order of Malta (SMOM), 47

Spain, colonial territory, 52–53, *54*, 55

Special Use Airspace, 169, 211n5

Spice Islands. *See* Molucca Islands

state boundaries (United States), 86, 96–101, 190n1

state law, variations in: administrative regulations, 86; taxation, 86; waterfront property, 26

Staten Island, NY, 101, *102*

St. Croix, 78

Stein, Mark, 98

St. George, UT, 103

St. Michaels, MD, 164

Stoeckl, Eduard, 76

strip chart, 166

subdivision map, 7

submarine topography, 74–76

Sunday Telegraph, 31

Sunni Arabs, 64

superdistricts, 113. *See also* multimember districts

Supreme Court: Ellis Island case, 101, *102*; on land-use zoning, 131–32; Oneida land claims case, 49–50; ruling on Tuskegee, AL gerrymander, 92; on sex-related businesses, 147; on voting rights, 105; on wetlands restrictions, 144

Surrey, BC, Canada, 36

survey. *See* land survey

Survey of India, 41

surveyor's notes, 19

Swains Island, 80–81

Syracuse, NY: municipal monument network, 21; redlining in, 123; truck-bridge collisions, 184

Syracuse Post-Standard, 126

Taj Mahal, 169

tax maps, 21–22, 29

temporary flight restrictions (TFRs), 164–65, 184

Terminus (Roman god of boundaries), 3

territorial sea, 71–73, 81–82

Texas, 95

thalweg, 100

Theater District (New York City), 135

Thirty-Eight State U.S.A., A (Pearcy), 97–98, *99*

Thurmont, MD, 164, *165*

tidal lands, 26

Time, 98

Times Square, 135, 151

Tiner, Ralph, 144

Tisdale, Elkanah, 108

title, real estate, 4, 6–7

title deed, 6, 22

Tito, Josip Broz (Marshal of Yugoslavia), 63

Tokelau, 80–81

topographic maps, content of, 3

Tordesillas, Treaty of, 52–53

Tordesillas Line, 52–53, *54*, 66, 193n4

town boundaries, 89–90

Town Government Belt, *89*, 90–91

township (element of U.S. Public Land Survey): areal unit, 14–15; horizontal reference, 15, *15*

township (minor civil division in Michigan, Pennsylvania, and several other states), 90

TransManche (cross-channel) Region, 31, *32*
transportation technology, as impetus for restrictive cartography, 2
Transylvania, 62
travel restrictions, diplomatic, 172–74
trespass, thwarted by property maps, 1–2
tripoint (in maritime boundaries), 78
truckers, navigation maps for, 160, 184–85
Turning Stone Resort and Casino, 49, 103
Tuskegee, AL, 92–94

underbounded cities, 92
underground facilities, 4, 23, 160–62
undersea cables, 84
Underwriting Manual (of Federal Housing Administration), 123, 129
United Arab Emirates, boundary with Saudi Arabia, 35
United Kingdom. *See* Britain
United Nations, 31, 58, 69, 70, 71, 74
United Nations Commission on the Limits of the Continental Shelf, 74–75
United Nations Convention on the Law of the Sea (UNCLOS), 73–74, 81–82
United States: air defense zones, 169; Antarctic claims rejected by, 68; exclusive economic zone, 73–74, 78–81; internal boundaries, 86–103; Libya's closing line rejected by, 72; no-fly zones over Iraq, 171; participation in Paris Peace Conference, 59–63; territorial sea, 72–73; travel restrictions for Russian diplomats, 172–74, 213n27
United States–Canada border: cleared area ("vista"), 35–37; maritime boundary, 75–77, 83, 197n21
United States–Mexico border: adjustment of, 38–39, *40*, 192n19; barrier construction along, 36–37; map symbols for, 3
urban growth, as impetus for restrictive cartography, 2
Uruguay, Antarctic aspirations, 68
U.S. Army, 166
U.S. Army Corps of Engineers, 144–45, 208n31
U.S. Coast and Geodetic Survey, 166

U.S. Coast Guard, 85
U.S. Customs and Border Protection, 37
U.S. Fish and Wildlife Service, 208n31
U.S. Geographic Board, 166
U.S. Geological Survey: boundary symbols, 87–88; cartographic treatment of Oneida Nation Territory, 48–49; reservation boundaries on, 48–49; *Topographic Instructions*, 49
U.S. Navy, 166, 167
U.S. Postal Service, 123, 166
USSR. *See* Russia
Utica, NY, 174

variance (in land-use zoning), 135–36
Vatican City, 46
vegetation line, 26
Venezuela, maritime boundary with United States, 70, 78–80
Vermont, 175
Victor (town), NY, 127
Virginia: land subdivision in, 13; travel restrictions for Russian diplomats, 213n27
Virginia City, NV, 137
Virgin Islands, 78
Voting Rights Act, 108

WalMart, 126
Washington (state), legal treatment of waterfront boundaries, 26
Washington, DC: on aeronautical charts, 162–63, 166–67, 168–70; and travel by Soviet diplomats, 173–74
Washington Monument, 162–63, 169, *170*
weighted voting in House of Representatives, 116
Welles, Sumner, 32, *33*
West Bank, 43–45
Western Territory (United States), subdivision of, 14
wetlands regulation, 3, 143–45
whaling, 64, 66–68, 80
White House, 162, *163*, 187
wildlife conservation, prohibitive cartography and, 2, 175, 177
Wilson, Woodrow, 59, 63
windshield survey, 22

wireless telephony: and locational restrictions on sex offenders, 158; location tracking and, 180

Wiscasset, ME, 137

World Court. *See* International Court of Justice

World Trade Center, 170, 187

World War I, 58, 166

World War II, 31, 58, 174

Wyoming, rectangular survey system for, 17

Yellow Sea, 72

Yemen, boundary with Saudi Arabia, 35

York, Duke of, 99

Yoruba, 58

Yugoslavia. *See* Jugoslavia

ZIP codes: acronym, 205n24; redlining with, 123–24, 129

zoning map, 1, 3, 130–43, 147–52